Hands-On Math Algebra

by

Pam Meader and Judy Storer

illustrated by
Julie Mazur

WALCH PUBLISHING

User's Guide
to
Walch Reproducible Books

As part of our general effort to provide educational materials that are as practical and economical as possible, we have designated this publication a "reproducible book." The designation means that the purchase of the book includes purchase of the right to limited reproduction of all pages on which this symbol appears:

Here is the basic Walch policy: We grant to individual purchasers of this book the right to make sufficient copies of reproducible pages for use by all students of a single teacher. This permission is limited to a single teacher and does not apply to entire schools or school systems, so institutions purchasing the book should pass the permission on to a single teacher. Copying of the book or its parts for resale is prohibited.

Any questions regarding this policy or request to purchase further reproduction rights should be addressed to:

Permissions Editor
J. Weston Walch, Publisher
P.O. Box 658
Portland, Maine 04104-0658

SUSTAINABLE FORESTRY INITIATIVE
Certified Chain of Custody
Promoting Sustainable Forest Management
www.sfiprogram.org

SGS-SFI/COC-US09/5501

1 2 3 4 5 6 7 8 9 10
ISBN 978-0-8251-6327-2

J. Weston Walch, Publisher
P.O. Box 658 • Portland, Maine 04104-0658
www.walch.com
Printed in the United States of America

Contents

Topic: Logarithms

Topic: Miscellaneous

To the Teacher

Since the early 1970s we have been teaching math to learners of all ages, from young children to adults, who represent many different cultures and socioeconomic backgrounds. We believe that all learners can do math by first overcoming any math anxiety and then participating in meaningful cooperative learning activities that foster the four major standards of Communication, Problem Solving, Connections, and Reasoning. These standards are founded in the Curriculum and Evaluation Standards for School Mathematics (1989), grades K–12. The draft of the content framework for *"Equipped for the Future" Standards for Adults* (1998) suggests that for adults to be "equipped for the future," they must be able to problem-solve, make decisions, and communicate effectively using math concepts and technology in an ever-changing world.

Our goal is to encourage all learners to "know math by doing math." To this end, we have developed activities called "labs" that enable the learner to discover mathematical concepts through a hands-on approach. Cooperative learning skills are developed through group activities in which each learner participates collaboratively as a team member. Communication skills are fostered through group discussion and written reactions to lab discoveries. Many of the labs are connected to real-life situations. Other labs require learners to form generalizations about mathematical revelations.

As teachers, we believe learning should be learner-centered, not teacher driven. The response from our learners has been favorable. As one student said, "Thank you for turning my math disability into a math ability."

— Pam and Judy

1. Candy Promotion

Teacher Page

Learning Outcome

Students will be able to:

- draw conclusions from tabulated data.
- use ratio and proportion for comparisons.
- become more informed on how advertising influences consumer spending.

Overview

Student teams will simulate market research groups. They will do a blind taste test of three kinds of Milky Way™ bars. They will use ratio and proportion to compare the fat and calorie content and serving sizes. They will also compare the wrappers of the candy bars to critically examine advertising claims and address consumer concerns.

Time

Two 45–60 minute periods

Team Size

3–4 students

Materials

Pieces of three different types of Milky Way™ bars for all the members of each team (regular Milky Way, 2.05 oz. or 58.1 g; Milky Way Dark, 1.76 oz. or 49.9 g; and Milky Way Lite, 1.57 oz. or 44.5 g); paper plates to display samples of candy; cutting utensil; calculators; overhead transparency of Master Candy Chart (page 3)

Procedure

Part One

1. Before beginning this lab, decide which bar will be labeled A, B, and C. Then cut the bars into samples, label the plates, and distribute them to each group. Each group should receive one sample of each candy bar for each team member. Retain the candy wrappers.
2. Put the master chart (page 3) on an overhead or copy it on the board so students can post their findings.
3. Do not reveal the product names until all students have logged their results. Once all students have finished tasting and posting the results, reveal which bar was type A, etc.
4. Allow time for the groups to discuss among themselves if taste alone is enough to promote the Milky Way Lite bar.
5. Open the discussion to the whole group once the smaller groups are finished.
6. Have students illustrate the results on the Master Chart with a visual display—a graph such as a circle, bar, or line graph—either by hand or on a computer.

Part Two

Distribute the candy bar wrappers for teams to examine the nutritional content of each bar. The following answers were based on the standard-size candy bars in the materials list above.

Answers

1. Milky Way regular **270**, Milky Way Lite **170**, Milky Way Dark **220**
2. Milky Way, 90/270 or **$33\frac{1}{3}$%** fat; Milky Way Lite, 50/170 or **29.4%** fat; Milky Way Dark, 70/220 or **31.8%** fat
3. a. Teams may approach this problem in many ways. They may wish to compare serving sizes by weight (grams or ounces) or by calories. They could convert all bars to the size of a Milky Way (regular) to see how large or how small a serving is. For example, the Lite bar is 44.5/58.1 of the serving size for a Milky Way regular or 76.6% of the serving. So the Lite bar's single serving is about 3/4 as large as that of the Milky Way regular.

   If teams choose to compare grams of fat to total grams, the following must occur: Convert sizes of bars to the Milky Way regular 58.1 g size and compare grams of fat.

   Lite bar $\dfrac{5\text{ g}}{44.5\text{ g}} = \dfrac{\mathbf{6.5}\text{ g}}{58.1\text{ g}}$ The Lite bar would have 6.5 g fat if it were the same size as the Milky Way regular.

   Dark bar $\dfrac{8\text{ g}}{49.9\text{ g}} = \dfrac{\mathbf{9.3}\text{ g}}{58.1\text{ g}}$ The dark Milky Way would have 9.3 g fat if it were the same size as the Milky Way regular.

   Teams will discover that the fat grams are not that different: Milky Way = 10 g, Lite = 6.5 g, and Dark = 9.3 g. While the Lite is definitely lower in fat, it is not 50% lower (see 3.b).

   Looking at the fat calories, since 1 g of fat is about 9 calories, the Lite bar now has 6.5 × 9 or **58.5** fat calories and the Dark has 9.3 × 9 or **83.7** fat calories, compared to the 90 fat calories of the Milky Way regular.

   Students may carry this further and look at calorie differences if all the bars were the same size.

   For the Lite bar $\dfrac{50\text{ fat calories}}{170\text{ total calories}} = \dfrac{58.5\text{ fat calories}}{\mathbf{199\ total\ calories}}$

   For the Dark bar $\dfrac{70\text{ fat calories}}{220\text{ total calories}} = \dfrac{83.7\text{ fat calories}}{\mathbf{263\ total\ calories}}$

   Therefore, if all bars were the same size, 58.1 g, the difference in calories between the Lite bar and the regular bar would be about 70 calories instead of 100 calories. The difference in calories for the dark bar and the regular bar would be only 7 calories, instead of 50 calories.

   b. This should be open for discussion. It gets students to scrutinize advertising claims. Their first reaction might be that the company lied. However, the company did not explicitly state which bars their Lite bar was compared to. It could have been bars with much higher fat content. Averaging also has an effect on the results.

   c. Packaging is also used to influence the customer to buy a product. Note the color of the wrappers. The Lite bar is wrapped in white, giving it the appearance of being "light". The dark bar has heavy, rich colors, indicating that this bar might even have more calories or a richer flavor.

Name ______________________________ Date ______________

1. Candy Promotion

Master Candy Chart

Team	Number of Team Members Choosing Candy Sample A:	B:	C:
1			
2			
3			
4			
5			
6			
7			
8			
9			
10			
	Total	Total	Total

(continued)

Name ______________________________ Date ______________

1. Candy Promotion *(continued)*

Part One

You and your partners work for the candy company that sells Milky Way bars. You have been asked to increase sales of the Milky Way Lite™ bar. First, you must conduct a taste test with the three different kinds of Milky Way bars on the market. You will use the results to determine how the Milky Way Lite bar's taste compares with the other two.

The Taste Test

You, acting as customers, will be given three samples of candy bars labeled A, B, and C. You will taste each, choose the one that tastes the best, and record your results. When your team is done, you will submit your results to the master board for comparison with other testers.

On the lines below, record how many in your group chose A, B, or C as the best-tasting candy bar. Then post these results on the master chart.

A ____________ B ____________ C ____________

After reviewing the results of the taste tests on the master chart, you will be told which letter represented the Milky Way Lite bar.

What conclusions can you draw about promoting the Milky Way Lite bar on taste alone?

____

____

____

____

____

How might you visually display the results of this survey? Make a graph that shows the results on the Master Candy Chart.

(continued)

Name ______________________________ Date ______________

1. Candy Promotion *(continued)*

Part Two

Many people are weight-conscious today. The U.S. government dietary guidelines recommend that no more than 30 percent of our daily calories come from fat. One way to manage this is to eat only foods that get no more than 30 percent of their calories from fat. There are about 9 calories in each gram of fat.

Read the nutritional information on the wrapper of each candy bar. Compare the nutrition facts for each 1-serving bar to see if nutrition is the way you want to promote the Milky Way Lite™ bar.

1. Compare the number of calories per serving.

   ______________, ______________, ______________
   Milky Way, regular   Milky Way Lite   Milky Way Dark

2. Fat calories represent what percent of each entire bar? Show your work below.

   ______________, ______________, ______________
   Milky Way, regular   Milky Way Lite   Milky Way Dark

3. As you begin your sales promotion and have people sample the Lite bars, questions arise that you must answer.

   (a) Each bar is sold as "one serving." One consumer argued that this description did not mean the same thing for all three bars. What did she mean by this? How could you explain it?

   ____

   ____

   (b) Previous advertising promoted the Lite bar by stating on the wrapper: **"170 calories 50% Less Fat Than Average of Leading Chocolate Brands."** Based on the results you found above, how could the company make this claim?

   ____

   ____

   (c) A third promotional idea might be to compare the wrappers of the three bars. What do you notice? How might this influence customers in making their selection?

   ____

   ____

Teacher Page

2. Estimating Wildlife Populations

Learning Outcome

Students will be able to:

- set up and solve proportions.
- work with inferential statistics and realize the importance of multiple trials to validate findings.

Overview

Teams will do a "wildlife study" using dried beans and a container to simulate animals and their habitat. After taking samplings comprising tagged and untagged animals and solving for proportions, each team will have an estimated average of the total number of animals. They will compare this to an initial guess and to the actual number of animals.

Time

1–2 class periods

Team Size

Pairs

Materials

Two kinds of dried beans for the creatures (white and red); shoe boxes, plastic freezer bags, or milk cartons for the habitat

Procedure

1. Have teams take three handfuls of red beans to be used as untagged creatures and place them in a container.
2. Then have each team take a handful of white beans (to be used as tagged creatures) and count them. Have them record this value, as it will stay constant throughout the lab.
3. Have each team add their handful of white beans (tagged creatures) to the red beans (untagged creatures) in the container.
4. Emphasize that each student on each team should estimate the total population.
5. You may have to review solving a proportion and make sure that students understand the letter notations used in the actual lab.
6. After students have worked through the lab, pull the class together for discussion using the questions that follow the lab as guidelines. It is interesting to see what students do when they gather no tagged creatures during a trial. Many times they choose to keep the data, but adding a zero greatly skews the results. Some students decide to throw that trial out and do another, which is the best choice, but this should be discovered through comparisons with other team members and class discussions. Students begin to see what error analysis is. They also begin to see problems that can skew the results. It's interesting to ask for these problems and see how many students identify.

Name ______________________________ Date ______________

2. Estimating Wildlife Populations

Purpose: One method of estimating a population is the **capture-recapture** technique. Naturalists often use this method to monitor the population of animals and fish. In this lab, you and your team will model this technique. Dried beans will represent creatures and a container will represent the habitat.

1. Have one team member grab three handfuls of **red** beans. Put these beans in a container (the **habitat**). These beans will represent **untagged creatures** in your habitat.
2. Have one team member grab a handful of **white** beans. Count the number of white beans. These beans will represent the number of **tagged creatures** (***T***) in your habitat. Record this total on the data sheet on page 9.
3. Take the habitat containing red beans. Put the white beans in with the red beans and mix the beans together. Guess the total number of beans in the container. Record your prediction on the data sheet on page 9.
4. To begin Trial 1, have one team member grab a handful of beans from the habitat. Separate the red beans (untagged creatures) from the white beans (tagged creatures). Count each kind. The number of white beans represents the number of **recaptured** tagged creatures, ***t***. The total number of red beans + the total number of white beans selected represents the **total number** of recaptured creatures, ***n***.
5. Record the number of recaptured tagged creatures (t) in column two of the data sheet. Then, record the number of total captured creatures (n) in column three of the data sheet.
6. Use the proportion below to estimate the total number of creatures in your team's habitat (***N***). T is **the constant**. It stays the same in each proportion. N is the unknown. In every trial sampling you will know T, t, and n. You must calculate N.

$$\frac{T \text{ (number of creatures tagged)}}{N \text{ (total number of creatures in the trial)}} = \frac{t \text{ (recaptured tagged creatures in trial)}}{n \text{ (total number of captured creatures in trial)}}$$

$$\frac{T}{N} = \frac{t}{n}$$

(continued)

Name ______________________ Date ______________

2. Estimating Wildlife Populations *(continued)*

7. Record the calculated value of N in column four of the data sheet.
8. Return the beans to the container.
9. Repeat this process to conduct nine more trials.

Sample trial results

If you have 58 tagged creatures (white beans) at the start, $T = 58$ in each trial. Next, if you have 132 total captured creatures in your handful in Trial 1, $n = 132$. If 24 of those 132 captured creatures are tagged creatures (white beans), then $t = 24$, or the number of recaptured tagged creatures.

$$\frac{\textbf{58}\text{ (Total number of tagged creatures)}}{\boldsymbol{N}\text{ (Total number of creatures in trial)}} = \frac{\textbf{24}\text{ (recaptured tagged creatures)}}{\textbf{132}\text{ (total number of captured creatures)}}$$

$$\frac{58(132)}{24} = N$$

$$319 = N \text{ (total number of creatures in trial)}$$

You and your team are now ready to conduct your ten trials to determine how many creatures are inhabiting your environment.

(continued)

Name ______________________________ Date ______________

2. Estimating Wildlife Populations *(continued)*

Capture/Recapture Data Sheet

Total number of tagged creatures (white beans), T.
This is the constant. ______________

Estimated number of all creatures in the habitat. ______________

Trials

Trial #	Number of recaptured tagged creatures (t)	Number of total captured creatures (n)	Find N, total number of creatures in trial $\frac{T}{N} = \frac{t}{n}$, $N=$
1			
2			
3			
4			
5			
6			
7			
8			
9			
10			
			Average: ______________

Now count the actual number of creatures in your team's habitat (plastic bag).

Red beans ________ White beans ________ Total ________

(continued)

Name ______________________________ Date ______________

2. Estimating Wildlife Populations *(continued)*

Capture/Recapture Inquiry Sheet

1. How does the average number from the ten trials compare to your estimated guess? ______________________________

____

____

____

2. Did you ever have *no* tagged creatures in your sample? ______________

Explain how this would affect the results. ______________________________

____

____

____

3. Do you think that the average of your ten estimates is a good estimate for the Total Number, N? Explain.

____

____

____

____

4. Do you think the population could be greater than any of your trials? ________

Explain. ______________________________

____

____

____

5. Compare the following:

(a) Actual number of total creatures – guess = ______________________________

(b) Actual number of total creatures – average from trials = ______________

Which is the closest to the actual, your guess or the trial average? ____________

____

(continued)

Name ______________________________ Date ______________

2. Estimating Wildlife Populations *(continued)*

Capture/Recapture Inquiry Sheet

6. Why is it important to return the beans to the container and mix the beans each time you repeat a trial? ______________________________

7. Why is it a good idea to base an estimate on several samples rather than just one sample? ______________________________

8. List some things that could happen to affect your results.

9. Do you think this is an accurate way to estimate populations? ______________

   Explain. ______________________________

3. The Solar System

Teacher Page

Learning Outcome

Students will be able to:

- convert various measures in different scales.
- compare the different sizes of the planets.
- appreciate the vast sizes and distances of the planets.

Overview

Student groups will scale down the measurements of the planets and their distances from the sun, then draw and cut out the planets to set up a scale model of the solar system.

Time

45–60 minutes

Team Size

2–3 students

Materials

Rulers, large sheets of paper, large space to display distances of planets (gym or outdoors), calculators

Procedure

For the first two activities, the students will be performing calculations using proportions. You might want to check to see if the students' proportions are set up correctly and if they are solving the proportions correctly.

Answers

Activity One	Activity Two
Mercury about $\frac{3}{4}$"	Diameter about $1\frac{1}{2}$"
Venus about $1\frac{1}{3}$"	Diameter about $3\frac{3}{4}$"
Earth about $1\frac{7}{8}$"	Diameter = 4"
Mars about $2\frac{7}{8}$"	Diameter about 2"
Jupiter about $9\frac{5}{8}$"	Diameter about $44\frac{1}{3}$"
Saturn about $17\frac{3}{4}$"	Diameter about $37\frac{1}{3}$"
Uranus about $35\frac{1}{4}$"	Diameter about $15\frac{3}{4}$"
Neptune about $55\frac{7}{8}$"	Diameter about 15"
Pluto about 73"	Diameter about 1"

The most difficult task for students is drawing circles with diameters of 37" and 44". Perhaps they could join hands in a circle with that diameter to see how large these two planets are compared to the other planets, or use a piece of string one-half the length of the diameter to draw a circle by taping one end of the string down and attaching a pencil or piece of chalk to the other end. The distancing of the planets might also prove challenging and will require room for the teams to spread out. Make sure they measure all distances from the point in the middle that will represent the sun.

Name ______________________________ Date ______________

3. The Solar System

In this lab, you will be making a scale model of the solar system. Do the calculations in Activities One and Two to find the correct measurements for your model.

Activity One

Given the scale 1 inch = 50 million miles, determine how far (in inches) each of the nine planets will be from your sun. Give your answers in approximate fractions.

1. Mercury, 36 million miles = ____________ inches
2. Venus, 67 million miles = ____________ inches
3. Earth, 93 million miles = ____________ inches
4. Mars, 142 million miles = ____________ inches
5. Jupiter, 484 million miles = ____________ inches
6. Saturn, 888 million miles = ____________ inches
7. Uranus, 1,764 billion miles = ____________ inches
8. Neptune, 2,791 billion miles = ____________ inches
9. Pluto, 3,654 billion miles = ____________ inches

Activity Two

Given a scale of 1 inch = 2,000 miles, find the approximate diameters of the planets in inches.

Mercury, 3,031 miles = ______ inches

Venus, 7,520 miles = ______ inches

Earth, 8,000 miles = ______ inches

Mars, 4,200 miles = ______ inches

Jupiter, 88,700 miles = ______ inches

Saturn, 74,600 miles = ______ inches

Uranus, 31,570 miles = ______ inches

Neptune, 30,200 miles = ______ inches

Pluto, 1,900 miles = ______ inches

(continued)

Name ______________________________ Date ______________

3. The Solar System *(continued)*

Activity Three

Using the values you found in Activities One and Two, try to construct a model of the solar system. Draw circles to represent each planet according to the diameter sizes you calculated. The larger planets may cause a problem. Discuss with your partner(s) how you might do this. Once you have drawn the planets and cut them out, place them according to the distances you had calculated in Activity One. Use the sun as the focal point.

Below, list some of the discoveries you have made about the solar system from this activity.

____

____

____

____

____

____

____

In making your model, you used very different scales for the diameters of the planets and the distances between planets. Why do you think it might be difficult to make a model using just one scale?

____

____

____

____

____

4. The Picture Frame Activity

Teacher Page

Learning Outcome

Students will be able to:

- recognize and describe number patterns.
- develop a formula.

Overview

Teams will estimate the number of tiles needed to build square "picture frames," build the frames, and then develop formulas for the number of tiles needed to build any square frame.

Time

Two 45–60 minute periods

Team Size

3–4 members

Materials

Per team—Approximately 50 square tiles (tiles may be made from oaktag or heavy paper if you don't have commercial tiles); centimeter grid paper for each member

Teacher—Overhead projector; overhead transparency of centimeter grid paper

Procedure

Day 1, Task One:

1. Set the stage by telling the students you are the production manager of The RIM Around Frame Company and you would like to hire them to be consultants for your company. Explain that they will work in teams; their job is to determine how many square tiles will be needed to produce various-sized picture frames.
2. Divide class into groups of three or four members and pass out lab sheets, grid paper, and tiles.
3. Allow 5–10 minutes for students to brainstorm the number of square tiles they will need to construct a 12-by-12 frame.
4. Instruct each group to build the frame, then make a scale drawing on cm-square paper.
5. Each team selects three more square frames of various sizes, builds them with the tiles, and draws a scale drawing on the grid sheet.

Day 1, Task Two:

1. Ask each team to analyze each of their constructions and scale drawings, then develop a method that will work for any size square frame imaginable.
2. Each team should select a spokesperson to share their ideas at the company meeting the following day.

(continued)

Day 2, Tasks Three, Four, Five:

1. Have a spokesperson from each team explain its method or strategy to the entire class.
2. Write each response on the board or overhead projector for all to view. Also reproduce one of the scale drawings that backs up their strategy. There are different ways to solve the problem. Here are the most common responses:
   - Multiply one side by 4 and subtract the 4 corners: $\mathbf{4 \bullet s - 4}$
   - Add the top and the bottom, then take away 2 from each side and add those also: $\mathbf{s + s + (s - 2) + (s - 2)}$
   - Add the first side and the next side minus 1, the third side minus 2, and the fourth side minus 3: $\mathbf{s + (s - 1) + (s - 2) + (s - 3)}$
   - One side minus 1, multiplied by 4: $\mathbf{(s - 1) \bullet 4}$
   - Area of large square minus area of inside square: $\mathbf{s^2 - (s - 2)^2}$
3. After all teams have presented their methods, discuss ways to write the procedures in “math shorthand”. Ask key questions:
   - What symbol could be used to replace the word “multiply”? (× or •)
   - What letter could stand for a side? (*s*), etc.

   The goal is to translate the written word into an algebraic expression. Ask, how is this like “one size fits all”? Explain that algebra is simply making a “one size fits all” strategy work for any similar problem.
4. Discuss how this formula is restricted to squares, but other formulas may be discovered to work for rectangles, pentagons, etc.

Extensions:

All of the formulas students derived can be reduced to one simple expression. Depending on the level of the class, you could discuss what strategy seemed the simplest and then show how the other strategies could be derived or simplified to look like the simplest strategy.

Name ______________________________ Date ______________

4. The Picture Frame Activity

The RIM Around Frame Company has hired your team as consultants. Your task is to determine how many square tiles you will need to buy in order to make various-sized square picture frames.

Task 1:

In your team, determine how many square tiles you would need to make a 12-by-12 frame.

We think we will need ______ tiles.

Take the number of tiles you need and build the frame. Did you select the correct number of tiles? If not, what do you think happened?

____

____

____

On your centimeter grid paper, make a scale drawing of your frame in which one centimeter square equals one tile.

Construct three more square frames of various sizes with your tiles. Then make scale drawings of your frames on the centimeter grid paper.

Task 2:

The production manager has asked you to develop a way to determine how many total tiles you will need to order for any size square frame constructed. Look at your scale drawings. Discuss ways that could be used to determine how many tiles are needed without building each frame and counting the number of tiles used. Describe your method or methods below.

____

____

____

____

(continued)

Name ______________________ Date ______________

4. The Picture Frame Activity *(continued)*

Task 3:

The production manager has learned that other companies have developed cost-saving methods of determining tile usage. There will be a company meeting in which each team will present its method. Be ready to support your team's method at the meeting. Write your presentation below.

Task 4:

After all presentations are over, choose the easiest method for determining tile usage and write it below.

Task 5:

In order to handle many orders, a computer programmer needs a "one size fits all" method for determining the number of tiles. Discuss with your boss a "math shorthand" way to describe your strategy and write it below.

(continued)

5. Integer Tiles

Teacher Page

Learning Outcome

Students will be able to:

- discover the properties for adding and subtracting integers.
- understand the concept of additive inverse.

Overview

Teams will use positive and negative integer tiles to perform addition and subtraction and to discover some properties of integers. Students will draw the tile operations on their worksheets and summarize their findings.

Time

1 hour

Team Size

Pairs

Materials

Groups of 20 tiles that are two different colors or that are marked with + and – symbols; integer mats (on page 24)

Procedure

Activity One

1. Pass out integer mats and positive and negative tiles for each pair of students. If tiles are not marked with + and – symbols, choose one color to represent negative numbers and one to represent positive numbers.
2. You may wish to model the activity, or you may have each pair work at its own pace.
3. The purpose of Activity One is to discover that adding a positive to a positive equals a positive.

Answers

1. [+] [+] [+]   [+] [+] [+] [+] [+]   8
2. [+] [+] [+] [+]   [+] [+]   6
3. [+]   [+] [+] [+] [+] [+] [+] [+]   8
4. positive integers
5. no
6. A positive plus a positive equals a positive.

Activity Two

Follow the same procedure as Activity One, only this time students should discover that a negative added to a negative is always a negative.

Answers

7. [–] [–] [–]   [–] [–]   –5
8. [–] [–] [–] [–]   [–] [–] [–]   –7
9. [–]   [–] [–] [–] [–] [–]   –6

10. negative integers
11. no
12. A negative plus a negative equals a negative.

Activity Three

1. In this activity, students should discover three possibilities when adding positive and negative integers. The answer will be negative, positive, or zero.
2. As students work with the tiles, have them line up positive tiles and negative tiles to form zero pairs.

Answers

13. 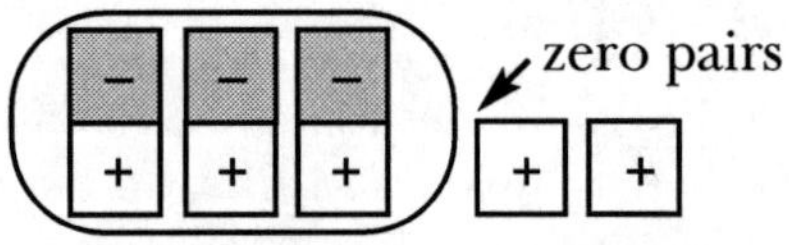

2

14. 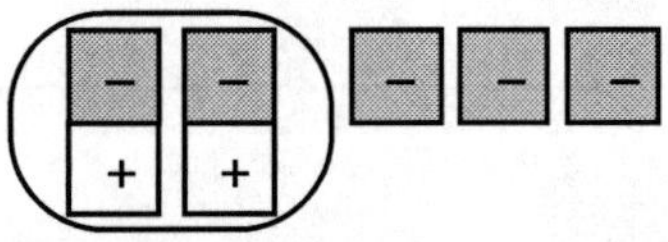
–3

15. 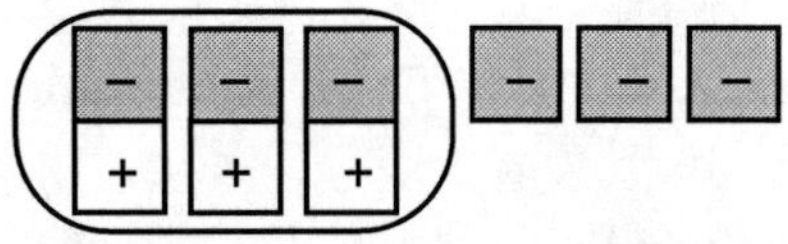
–3

16. 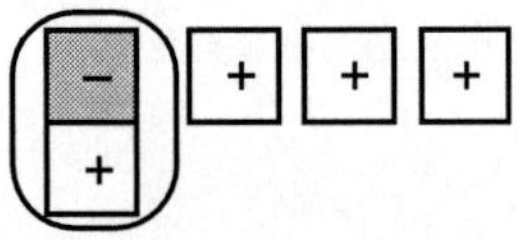
3

17. 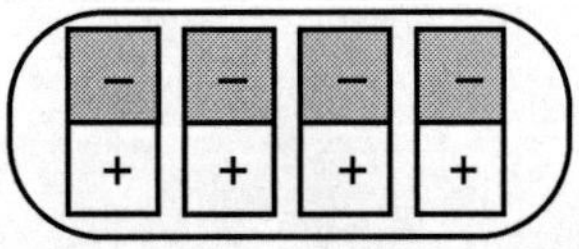
0

18. positive and negative integers
19. Possible answers might be: It seemed to be like subtracting. A greater number of negatives meant a negative answer; a greater number of positives meant a positive answer; if negatives and positives were equal, then the answer was zero.
20. When adding positive and negative integers, the sign of greater amount, whether it is positive or negative, will be the sign of the answer. Subtract the smaller amount of tiles from the larger to get the answer.

For the summation, you may wish to share each group's rules or have each group post its rules on the board.

Name ______________________________ Date ____________________

5. Integer Tiles

Activity One: Adding Integers

We are going to be using positive and negative tiles to discover some properties of adding and subtracting. Each positive tile has a value of 1. To show the number positive 3, you would place 3 positive tiles on the mat.

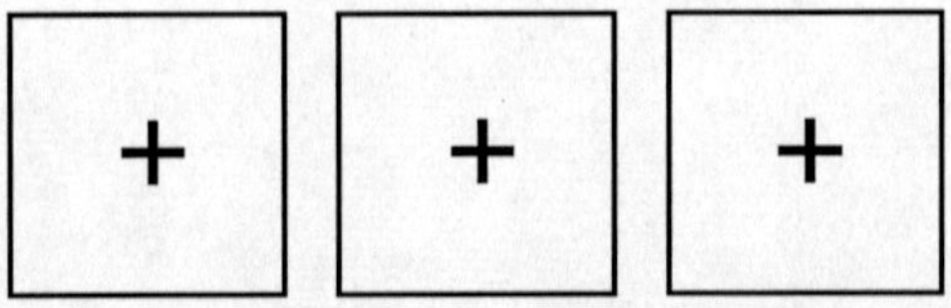

= 3

To show negative integers you will use the negative tiles. Each negative tile has a value of negative 1. To show the number negative 4, you would place 4 negative tiles on the integer mat.

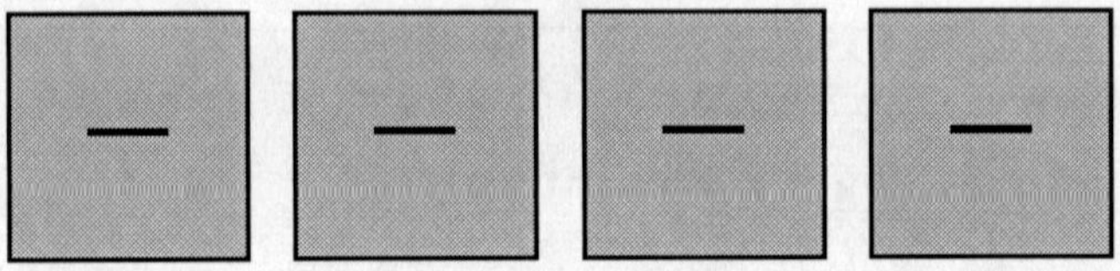

= –4

Try the following additions with the tiles and the integer mat.

	Draw the tiles here:	**Answer:**
1. 3 + 5		______
2. 4 + 2		______
3. 1 + 7		______

4. In every example above, what type of numbers were you adding?

____

5. Did you ever use a negative tile? __________

6. What rule can you develop for adding positive integers? ____________________

____

(continued)

Name ______________________________ Date ______________

5. Integer Tiles *(continued)*

Activity Two: Adding Negative Integers

This time we will work with the negative tiles. Show the following addition problems with your tiles.

	Draw the tiles here:	**Answer:**
7. –3 + –2		________
8. –4 + –3		________
9. –1 + –5		________

10. In every example, what type of numbers were you adding?

____

11. Did you ever use a positive tile? __________

12. What rule can you develop for adding negative integers? __________

____

____

Activity Three: The Zero Pair

When two tiles of unlike signs are combined, they form a zero pair. That is, their value = zero.

Example: 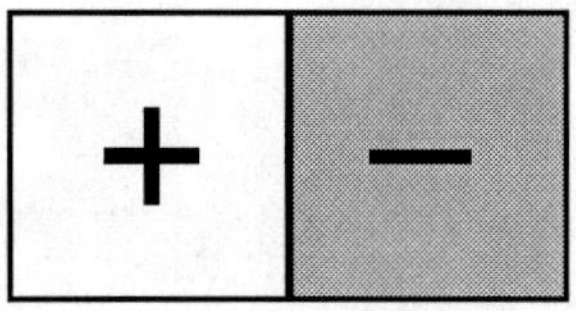

= 0

When you put the two together, they form a zero pair and cancel each other out.

When you combine positive and negative tiles, find how many form zero pairs and remove them from the mat. The tiles that remain show the answer.

Show the following additions with the tiles. Remove all zero pairs and find the answer.

(continued)

Name ______________________ Date ______________

5. Integer Tiles *(continued)*

Draw the tiles here: **Answer:**

13. $-3 + 5$ ________

14. $-5 + 2$ ________

15. $-6 + 3$ ________

16. $4 + -1$ ________

17. $-4 + 4$ ________

18. In every example, what types of numbers were you adding?

____

19. What did you notice happening each time you added the integers with different signs?

____

____

20. What rule(s) could you develop for adding positive and negative integers?

____

____

Summation:

A positive + a positive = ______________________

A negative + a negative = ______________________

A negative + a positive can be __________, __________, or __________

because ______________________________

(continued)

Name ______________________ Date ______________

5. Integer Tiles *(continued)*

Integer Mat

(continued)

6. Subtraction with Integer Tiles

Teacher Page

Learning Outcome

Students will be able to:

- visualize the process of subtraction with positive and negative integers.
- develop properties for subtracting with integers.

Overview

Student teams will arrange tiles or chips of two colors to illustrate subtraction of integers, and then draw the tile operations on their worksheets. Students will begin to devise a plan for subtracting any integer.

Time

30 minutes

Team Size

Pairs

Materials

Tiles of two different colors or that are marked with + and – symbols; integer mats (page 24)

Procedure

Part One

1. Pass out integer mats and tiles of two different colors. If tiles are not marked with + and – symbols, designate one color to represent positive numbers and one color to represent negative numbers.
2. Demonstrate the first activity on the student page.
3. As students try the activities, make sure that they draw the subtraction they have illustrated with the tiles.

Answers

1. 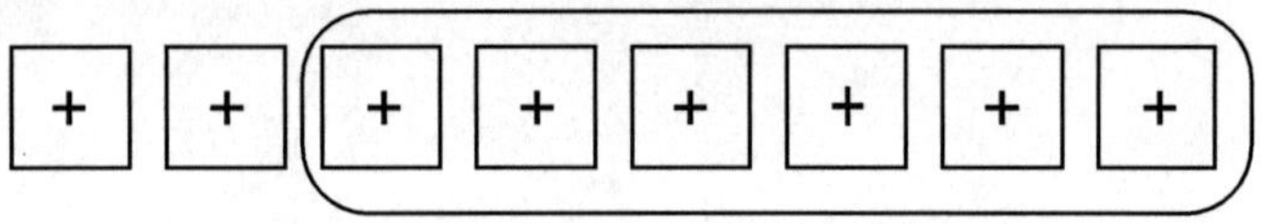

= 2

2. 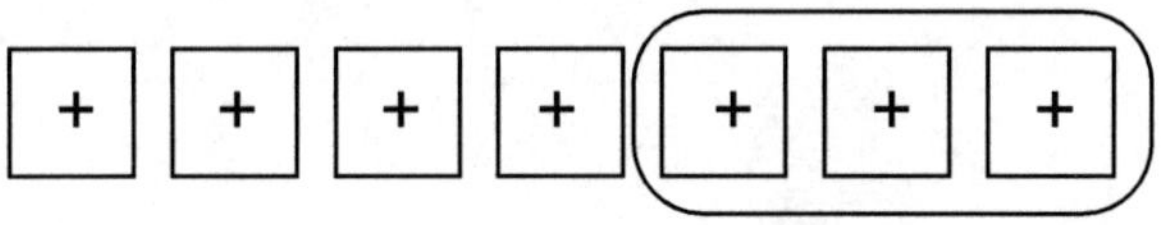

= 4

(continued)

Part Two

1. The second part of the lab illustrates subtracting a negative number with integer tiles. This section may prove difficult for students, and therefore it is very important that you demonstrate the first example, 3 – 8.

The last questions require the groups to devise a plan to subtract any integer. Discuss the rule that "subtracting an integer is the same as adding the opposite" and show how it is illustrated with the tiles.

Answers Number 3 is shown on the student page.

4. 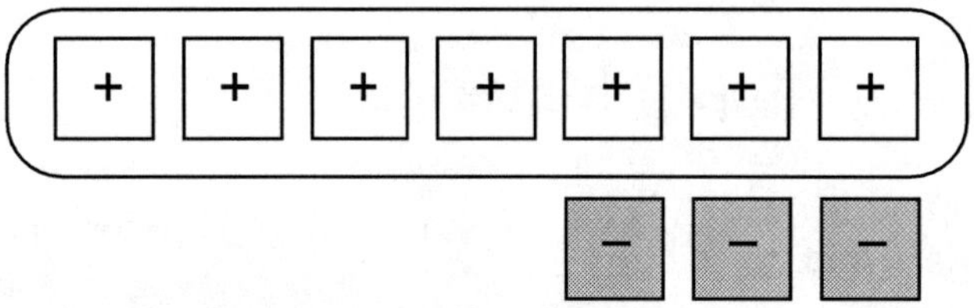

= –3

5. 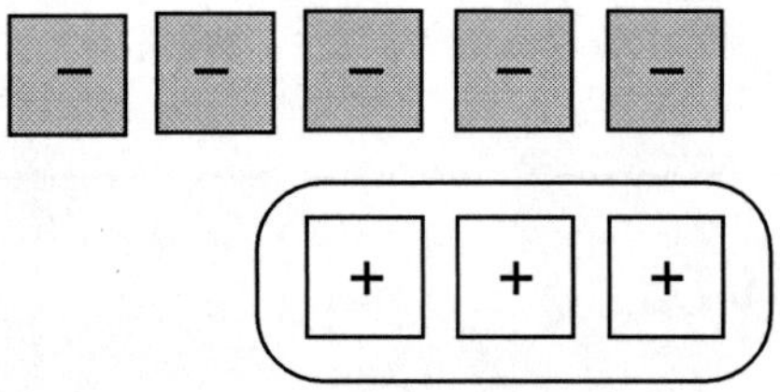

= –5

6. This last problem presents the most difficulty for students. Have them look at what is being subtracted to see if they need to add any zero pairs.

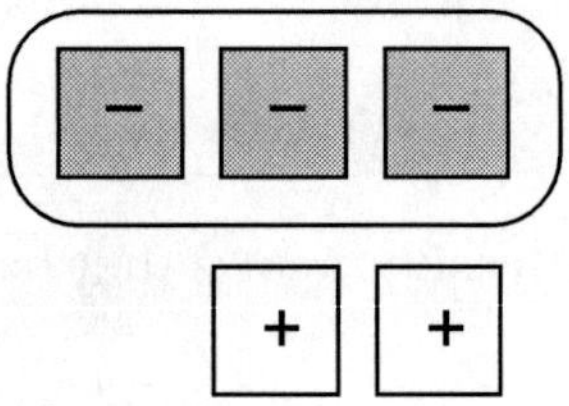

= 2

Name ______________________________ Date ______________

6. Subtraction with Integer Tiles

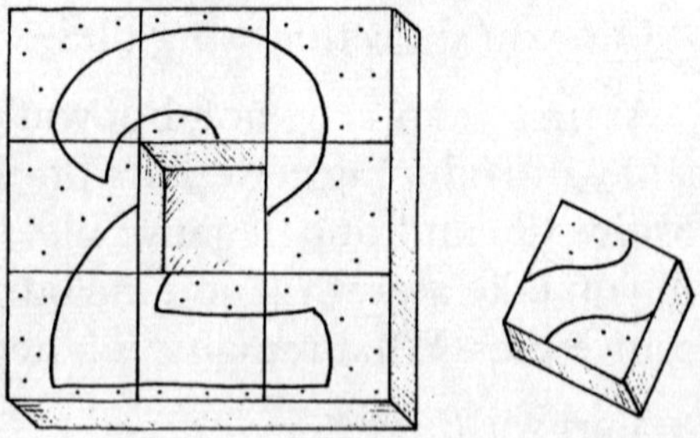

Part One

When you subtract with tiles, you remove the tiles that are being subtracted. What remains is the answer.

Example: To show 7 – 4 with tiles, you would start with 7 tiles on the mat.

Remove 4 tiles from the mat. In the drawing below, we circled the tiles that are to be removed:

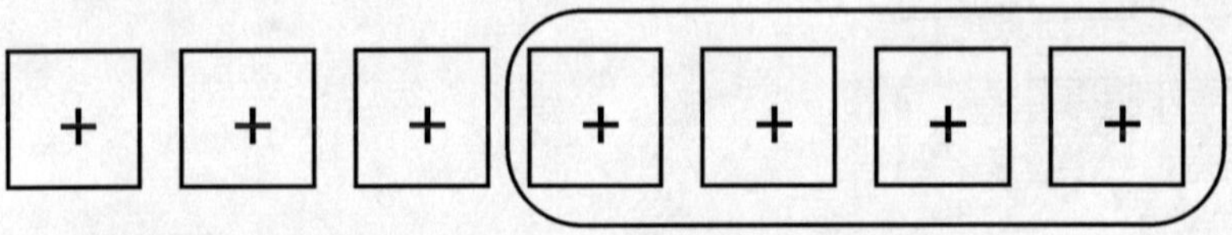

The answer is 3:

Show the following subtractions with the tiles. Draw the tiles beside each problem.

Answer:

1. 8 – 6 ________

2. 7 – 3 ________

Part Two

Using tiles, can you do the following?

3. 3 – 8

(continued)

Name ______________________ Date ______________

6. Subtraction with Integer Tiles *(continued)*

Can you show this using tiles?

At first, it appears not, but you can subtract more than you now have on the integer mat by introducing **zero pairs** until you have enough. (Each zero pair is made up of one positive tile and one negative tile.) In this example you start with 3 tiles on the mat. How can you take away 8? If you introduce 5 zero pairs, you will have enough tiles to remove 8 positive tiles. What remains is 5 negative tiles, which is the answer.

Start with:

You can't complete the subtraction until you introduce 5 zero pairs:

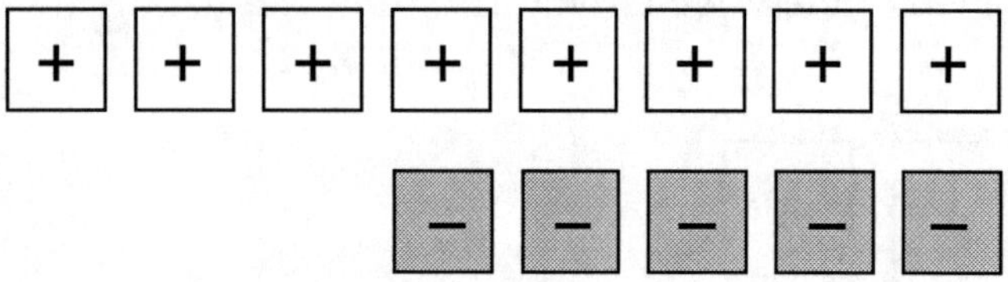

Now you can remove the 8 positive tiles. This leaves you with 5 negative tiles, which is the answer.

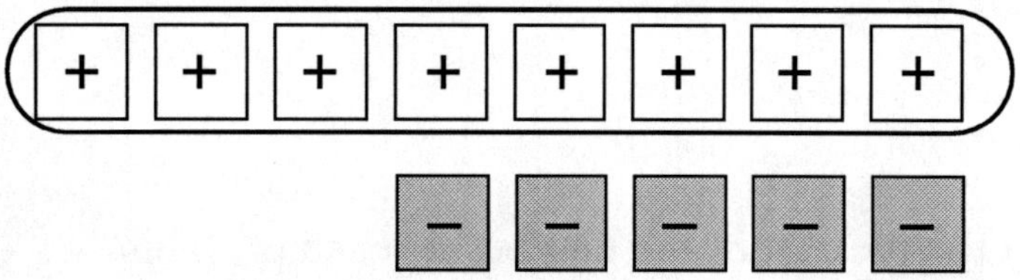

Try the following subtraction using tiles.

	Draw the tiles here:	**Answer:**
4. 4 – 7		______
5. –2 – 3		______
6. –1 – (–3)		______

What do you notice about subtracting integers? ______________________

____

Are there any rules you can see? ______________________

____

Teacher Page

7. Get It in Gear Game

Learning Outcome

Students will be able to:

- perform addition and subtraction with integers.
- distinguish between an operational sign and a sign (positive or negative) for a number.

Time

30 minutes or more

Team Size

Two students per team; two teams for each number line

Overview

Students themselves will act as game pieces, moving back and forth on life-size number lines to perform operations with positive and negative integers.

Materials

Number lines taped to or drawn on the floor (one number line for every four students); cards that have operations with integers on them (see page 31)

Procedure

This activity would work best in a large area like a gym, or even outside, if feasible. This activity definitely appeals to the kinesthetic learner. The activity also clearly delineates between a sign signaling an operation (**facing**) and a sign of a number (**moving**). The activity provides clear images about *why* –(–) results in + and also *why* –(+) and +(–) have the same outcome.

1. Photocopy and cut out cards so each team has a selection to pick from.
2. Draw number lines (–10 through +10) with chalk or mark them with masking tape. You could have each team make their own number line first to reinforce the order of the integers on the number line.
3. Select the teams and then match up two teams for each number line; teams will take turns acting and observing.
4. Make sure the teams understand the rules and how to move on the number line. It might be beneficial to model an addition or subtraction example so that students see the difference between movement for addition and subtraction as well as direction change mandated by the signs of the numbers.

   Example: The example 3 + (–5) would be acted out as follows: The first number is 3, so the student would stand on 3 on the number line. Addition means ***face forward***, which would mean to the right or facing the front, depending on how the number line is oriented. The negative sign inside the parentheses is the ***moving command.*** In this example **–5** means ***move backward*** 5 spaces. The student should land on –2 for the answer.

(continued)

Example: (–2) – (+3)
The first number tells you to start at –2 on the number line. The subtraction sign means ***facing the back (or left)***. The positive sign inside the next parentheses means ***move forward*** 3 spaces (not forward in the room, but forward in the direction you're facing). Since you are already facing in a direction, you continue in that direction. The answer will be –5.

5. Once students are clear on how to act out each number phrase, have teams pick a card and act out the phrase on each card. At the end of each round, teams on each number line switch between acting and observing. The first team to earn 5 points wins.

Name ________________________________ Date ________________

7. Get It in Gear Game

Directions: Two teams will be placed by a number line on the floor. One team will act out the mathematical phrase while the other team observes and determines if the team is right. One point is awarded when a team acts out a phrase correctly. If the team acting out the phrase does it incorrectly, the observing team receives the point. The first team to receive 5 points wins the game. Each round, the teams switch roles. In one round they are the observers; the next round they act out the phrase.

How to play the game

1. Determine which team will be the observers for Round 1 and which will be acting out the phrase. In Round 2, and thereafter, the teams reverse roles.
2. Operation symbols for adding (+) and subtracting (–) are called the **facing** commands. Addition means you face **forward (right)** while subtraction means you face to the **back (left)**.
3. Symbols inside the parentheses are called the **moving** commands. The positive sign means **move forward** while the negative sign means **move to the back.**
4. The observing team draws a card from the deck and reads it to the acting team. The acting team decides how to move on the number line and finds the answer. If the answer is correct, the acting team receives 1 point. If it is not correct, the observers receive the point.
5. The first number in the phrase is where the acting team starts its player on the number line.
6. Reverse roles for Round 2, etc.
7. The first team to earn 5 points wins.

Scoreboard

	Round 1	**Round 2**	**Round 3**	**Round 4**	**Round 5**
Team 1					
Team 2					

(continued)

Name ______________________________ Date ______________

Activity Cards for Get It in Gear Game

(+4) + (–3)	(–2) + (–3)	(–5) + (+4)
(+4) – (–3)	(–5) – (+3)	(–3) – (–3)
(+3) + (–3)	(6) + (–3)	(–4) + (–3)
(–3) + (–4)	(–2) – (–3)	(–2) – (+3)
(+5) + (–3)	0 + (–2)	(+2) + (–3)
(+6) – (–2)	(–2) – (0)	(–4) – (–3)
(+5) + (+3)	(+7) + (–3)	(–7) + (–3)
(–2) + (–3) – (+4)	(+5) – (–3) + (–4)	(–2) – (–3) – (+4)

Teacher Page

8. Powers of Ten

Learning Outcome

Students will be able to:

- discover the rules for multiplying and dividing with powers of 10.
- use the correct calculator key for powers of 10.
- work with positive and negative powers of 10.

Overview

Pairs of students will use a scientific calculator to practice multiplying and dividing with positive and negative powers of ten. Students will discover and write out the rules for these operations.

Time

45–60 minutes

Team Size

Pairs

Materials

Scientific calculators

Procedure

Before beginning the lab, show students how to use the EXP or EE key. Note that entry for powers of 10 may be different for various calculators. Usually if the calculator has an EXP key, then hitting the EXP followed by the exponent and the equals sign will give the required value. If the calculator has an EE key, it may be necessary to key 1 instead of 10, then EE, then the power, followed by the equals sign, then convert to standard notation by pressing the INV key followed by the EE key. Make sure students are finding answers in standard notation, not scientific notation.

Answers

Activity One

1. 10
2. 100
3. 1,000
4. 10,000
5. 100,000
6. 10
7. 100
8. 1,000
9. 10,000
10. 100,000

Students should notice that 10×10 is the same as 10^2, and so forth. They should also see that the number of zeros in multiples of 10 corresponds to the exponent value in the power of 10.

(continued)

Activity Two

1. 3456.7
2. 2.34
3. 34,500
4. 12,000
5. 102
6. 101
7. 103
8. 104
9. 0.76589
10. 0.009422
11. 0.000878
12. 0.0045678

Students should see that multiplying by powers of 10 results in movement of the decimal point. The decimal point moves one place for each zero in the power of 10, or the value of the exponent.

Activity Three

1. 0.076589
2. 3.4567
3. 0.00125667

Students should see that the results of multiplying with negative exponents are the opposite of multiplying with positive exponents; the decimal point moves to the left.

4. 76589.00
5. 34567.0
6. 125667.0

Students should see that when you divide by positive powers of 10, the decimal point moves to the left, but when you divide by negative powers of 10, the decimal point moves to the right.

Name ______________________________ Date ________________

8. Powers of Ten

Activity One

Compute the following on your calculator.

1. $1 \times 10 =$ _____
2. $10 \times 10 =$ _____
3. $10 \times 10 \times 10 =$ _____
4. $10 \times 10 \times 10 \times 10 =$ _____
5. $10 \times 10 \times 10 \times 10 \times 10 =$ _____

Using the EXP or EE key on your calculator, find the following:

6. $10^1 =$ _____
7. $10^2 =$ _____
8. $10^3 =$ _____
9. $10^4 =$ _____
10. $10^5 =$ _____

What do you notice about the answers to 1–5 and the answers to 6–10?

____

____

____

____

____

What rule(s) can you make about powers of 10 from your observations above?

____

____

____

Activity Two

With your partner and a calculator, try the following examples.

1. $34.567 \times 100 =$ _____________
2. $0.234 \times 10 =$ _____________
3. $34.5 \times 1{,}000 =$ _____________
4. $1.2 \times 10{,}000 =$ _____________

(continued)

Name ______________________ Date ______________

8. Powers of Ten *(continued)*

With your partner, discuss any patterns you see. Summarize them below.

____

____

____

____

Rewrite the examples on the previous page using powers of 10.

5. $34.567 \times 100 = 34.567 \times 10^{?} =$ ______________
6. $0.234 \times 10 = 0.234 \times 10^{?} =$ ______________
7. $34.5 \times 1{,}000 = 34.5 \times 10^{?} =$ ______________
8. $1.2 \times 10{,}000 = 1.2 \times 10^{?} =$ ______________

What do you notice about the exponent and the number of decimal places moved in the answer?

____

____

Use the rules that you discovered to multiply the following example without a calculator.

$$3.456 \times 10^2 = \_\_\_\_\_\_\_\_\_\_$$

Write a **generalization** for multiplying numbers by powers of 10 with positive exponents:

____

____

____

What do you think will happen when you divide by powers of 10? Write your prediction below.

____

____

Try the following examples with your calculator and see if your prediction is correct.

9. $76.589 \div 100 =$ ______________
10. $9.422 \div 1{,}000 =$ ______________
11. $0.878 \div 10^3 =$ ______________
12. $45.678 \div 10^4 =$ ______________

(continued)

Name______________________________ Date ______________

8. Powers of Ten *(continued)*

Was your prediction correct? Write a **generalization** for dividing a number by a power of 10 with positive exponents:

____

____

____

Activity Three

Enter the following into your calculator. Write the answers below.

1. 76.589×10^{-3} ______________
2. 345.67×10^{-2} ______________
3. 12.5667×10^{-4} ______________

What do you notice about the exponent and the number of decimal places moved in the answer? In which direction does the decimal point move?

____

____

Write a rule for multiplying by powers of 10 with negative exponents:

____

____

____

What do you think will happen if you divide by powers of 10 with negative exponents?

____

____

Try the following examples to see if you are correct.

4. $76.589 \div 10^{-3}$ ______________
5. $345.67 \div 10^{-2}$ ______________
6. $12.5667 \div 10^{-4}$ ______________

Write a rule for dividing by powers of 10 with negative exponents:

____

____

____

9. Distributive Property

Learning Outcome

Students will be able to:

- through patterns, discover the distributive property.
- through area, understand that the greatest factor is the common edge that rectangles share.

Overview

Student pairs will draw rectangles with common sides, then find their areas. As they examine more such rectangles, they will uncover the principles of the distributive property. They will also think about and describe the use of the perimeter formula and the distributive property in a real-life situation.

Time

30–40 minutes

Team Size

Pairs

Materials

Centimeter grid paper; scissors

Procedure

Activity One

1. Pass out lab sheet, scissors, and centimeter grid paper to each team.
2. Stress that the two rectangles formed from the larger rectangle do not have to be identical. However, students must divide the larger rectangle into two smaller rectangles, not triangles. Some possibilities are:

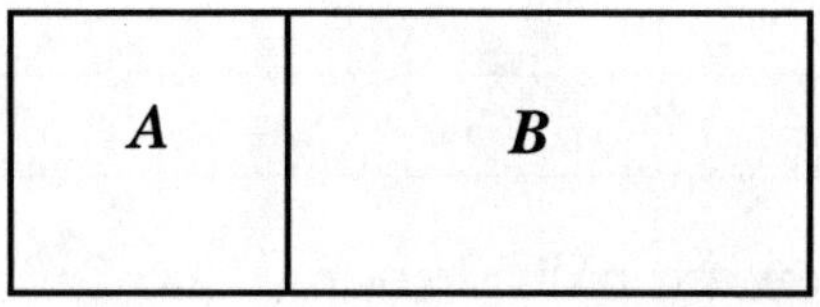

or

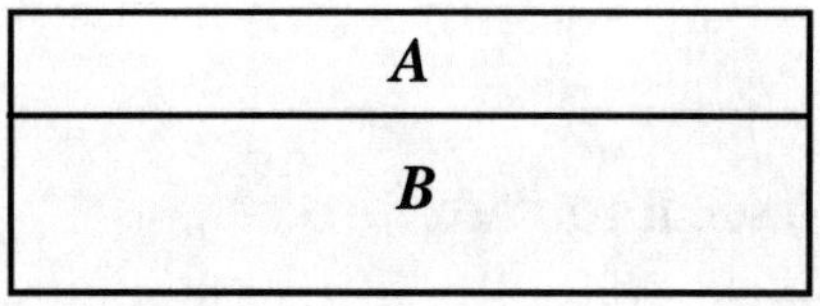

3. As students post their dimensions on the board, two patterns should appear. First, there will always be a common edge.
   For example, in the first diagram, suppose the rectangles had the following dimensions:

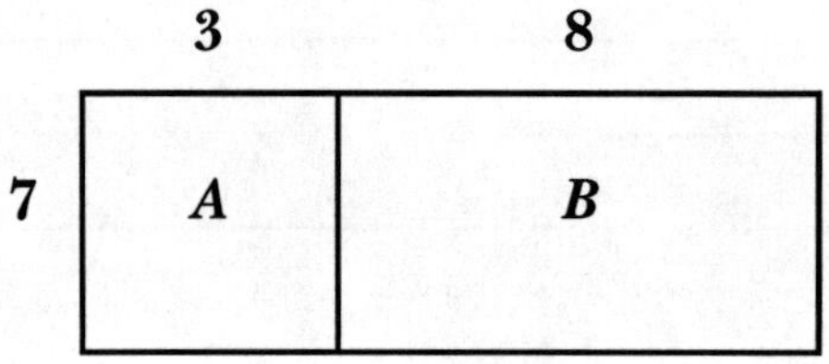

(continued)

The posting on the board would be: $7 \times 3 + 7 \times 8 = 7 \times 11$. Here the common edge is the measurement all three rectangles share, which is 7.

In the other example, given the following dimensions,

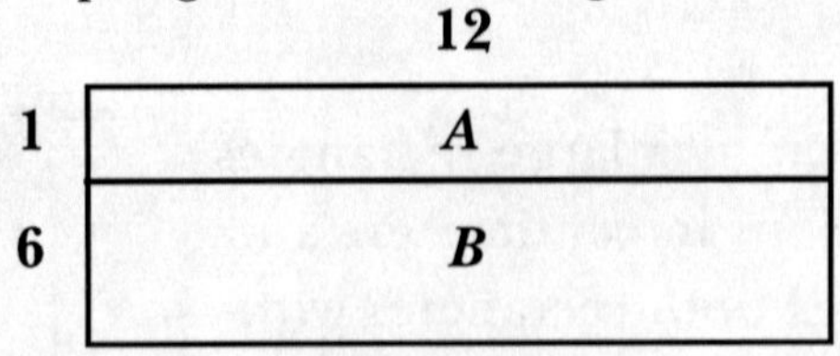

the common edge would be the length, or 12. The posting on the board would be: $1 \times 12 + 6 \times 12 = 7 \times 12$.

4. Once students have shared their discovery of common edges, circle the common edge in every example posted on the board to "prove" that it happened in each one.
5. The second discovery the students should make is that the other (not the common) dimensions of the smaller rectangle add up to the second dimension of the large rectangle. For example, in the first illustration, $8 + 3 = 11$; in the second illustration, $1 + 6 = 7$.
6. Post some more concrete examples to illustrate these discoveries. Ask, "What would $2 \times 3 + 2 \times 5$ equal?" (etc.), always showing that the common factor is the common edge of the rectangles.
7. Ask "What would $a \times b + a \times c$ equal?" Hopefully after the many concrete examples above, students will be able to state the abstract formula for the distributive property.
8. Continue with this procedure, starting with the larger rectangle. Ask, "What would $3(4 + 3)$ equal?" Make sure students use the distributive property and do not simply add 4 and 3 and multiply by 3. Do several concrete examples, then move to the abstract.
9. Ask, "What would $a(b + c)$ equal?" When you have finished, let students know that they have discovered the distributive property.

Activity Two

This activity can be used as an extension. If students really grasp the distributive property, they should be able to see that the perimeter formula for a rectangle is another example of the distributive property.

$$P = 2L + 2W = 2\ (L + W)$$

Students should be able to explain, for example, how figuring out fencing for a rectangular yard or determining the wallpaper border necessary to go around a rectangular room are real-life examples of the distributive property.

Name ______________________________ Date ________________

9. Distributive Property

Activity One

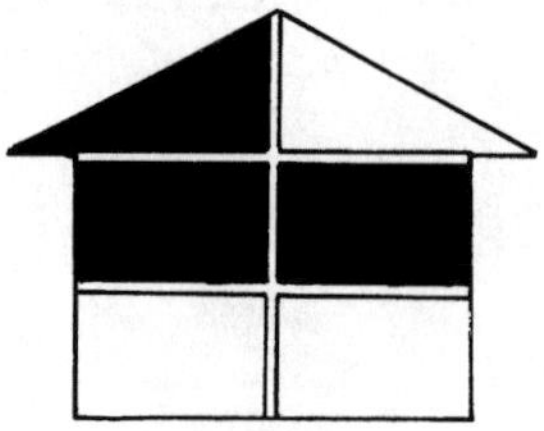

1. On centimeter grid paper, cut out two large rectangles, the same size. Find the dimensions in **centimeters** and record on your rectangles. Label both rectangles with a large *C*.

   The dimensions of my rectangle are length: __________ width: __________

   Next determine the **area** of rectangle *C* and record here.

   ____

   area of rectangle

2. Take only one of the rectangles and cut it into two smaller rectangles of **different sizes**. Label one rectangle *A* and the other *B*. Find the dimensions of each of your rectangles, label, and record here.

   Rectangle *A*'s dimensions: ______________________

   Rectangle *B*'s dimensions: ______________________

   Next determine the area of each rectangle and record.

   Area of *A* ____________, Area of *B* ____________

3. Do the sum of the areas of *A* and *B* equal the area of the large rectangle you marked *C*? __________

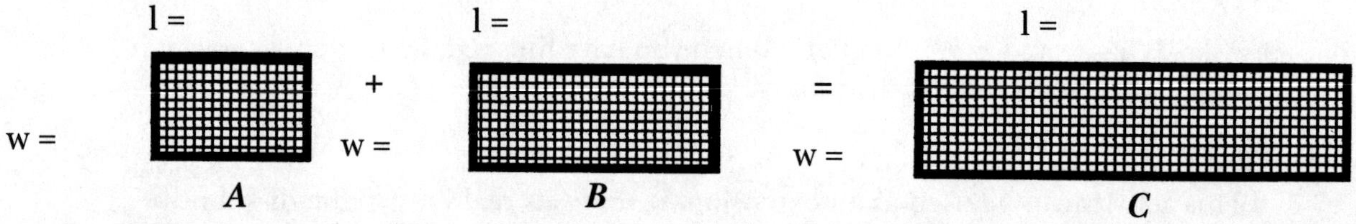

__________ + __________ = __________

$l \times w$ $l \times w$ $l \times w$

4. Next take rectangles *A* and *B* and place them together to equal the area of *C*. Label the dimensions of each rectangle below.

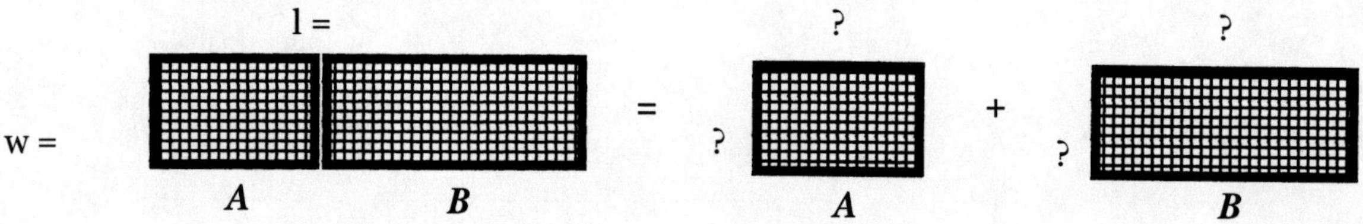

(continued)

Name ______________________________ Date ______________

9. Distributive Property *(continued)*

5. You are now going to post the dimensions of each of your rectangles on the board. Also record all data in the chart below.

Class Dimensions			
Name	Rectangle A l x w	Rectangle B l x w	Rectangle C l x w

(continued)

Name ______________________ Date ______________

9. Distributive Property *(continued)*

6. With your partners study the data and discuss any patterns you observe. List below and be prepared to discuss your findings with the class.

____

____

____

____

____

7. After class discussions, can you state the generalization for the distributive property for any rectangle using the diagram below?

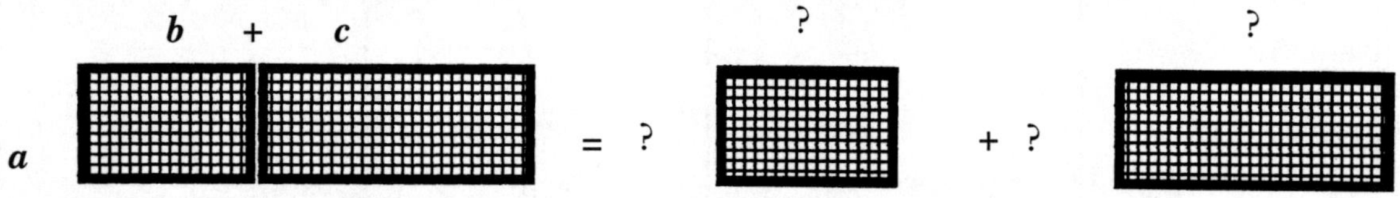

The distributive property states:

____

____

Activity Two

Is the perimeter formula for a rectangle $P = 2(L + W)$ an example of the distributive property? Think of a real-life situation in which you would need the perimeter formula. Then prove that your situation could be solved using the distributive property and visuals or demonstrations.

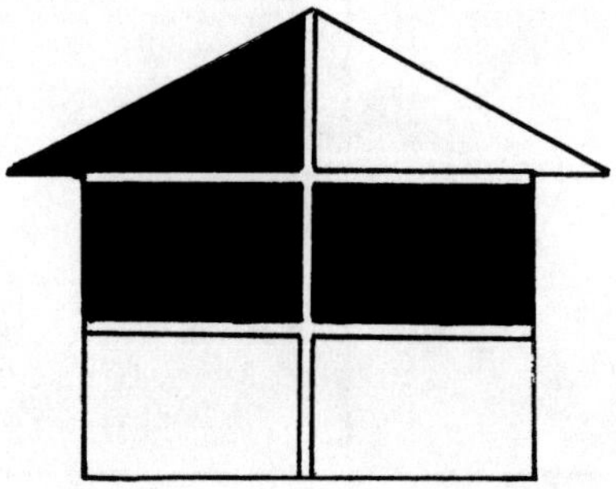

10. Discovering Algebra Tiles

Learning Outcome

Students will be able to:

- recognize the physical model for variables and constants.
- combine like terms both physically and algebraically.

Overview

Teams will examine algebra tiles and learn to combine like terms with them.

Time

30 minutes

Team Size

Pairs

Materials

Algebra tiles that include x^2-tiles, x-tiles, and constant tiles. Each team should have at least 5 x^2-tiles, 10 x-tiles, and 10 constants.

Procedure

Activity One

1. This activity is designed to acquaint students with algebra tiles. A variety of responses should come from the first part. Possible answers are: There are square tiles and rectangular pieces. The tiles are different colors (depends on set used; some sets have positive and negative tiles that are different colors, while other sets have the variable tiles in one color and the constants in another color). Some of the tiles seem to fit together while others do not. Please note this for students, as it will become more important when the students use the tiles to factor.
2. In the second part of Activity One, it is important to have students trace the different tiles to provide a spatial sense of each tile and to notice dimensions. For example, an x^2-tile has dimensions x by x, an x-tile has dimensions 1 by x, and a unit tile has dimensions 1 by 1.

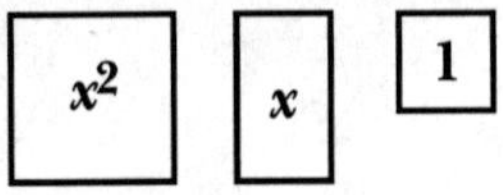

After they have completed their tracings, assist the students in labeling the tiles.

Activity Two

1. The purpose of this activity is for the student to see how like terms look.
2. If students put squared tiles together, x-tiles together, and unit tiles together, they probably won't then algebraically combine $2x^2$ with $4x$ and call it $6x^3$.

(continued)

Answers

1. 
2. 
3. 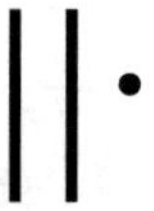
4. 
5. 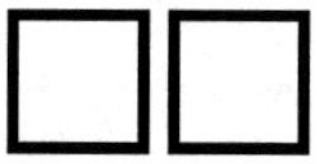
6. 
7. 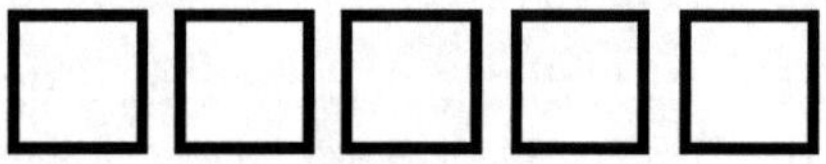 $= 5x^2$
8. 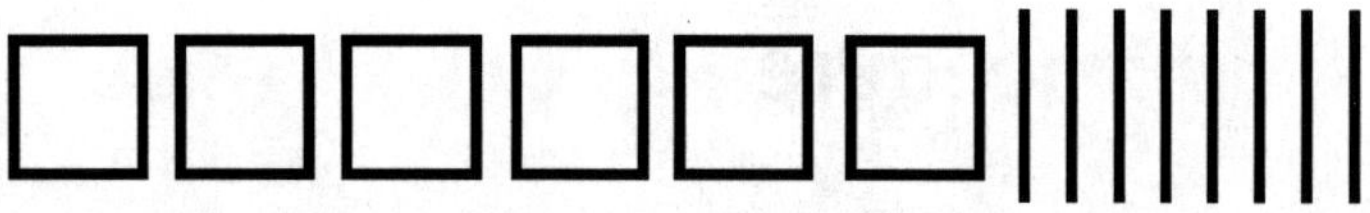 $= 6x^2 + 8x$
9. 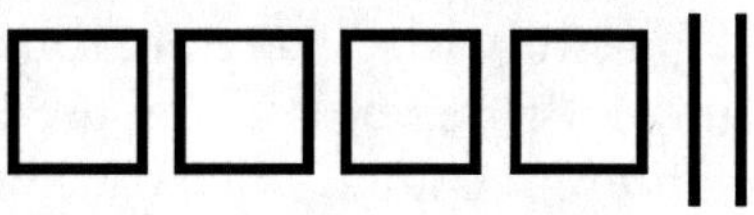 $= 4x^2 + 2x$
10.  $= 3x^2 + 6x + 6$

Name ______________________ Date ______________

10. Discovering Algebra Tiles

Activity One

With your partner, dump all your tiles onto the table. In the space below, jot down all the things you notice about the algebra tiles.

Trace each kind of tile below. We will label them together.

(continued)

Name ______________________________ Date ______________

10. Discovering Algebra Tiles *(continued)*

Activity Two

With your tiles, show the following. Draw the tiles below each example. Let □ represent x^2, | represent x, and • represent units.

Example: $x^2 + 3x + 2$ would look like □ | | | • •

1. $x + 1$

2. $x + x$

3. $2x + 1$

4. x^2

5. $2x^2$

6. $3x^2 + 4x + 3$

With your tiles show the following additions. Then, for each example, group "like tiles" together and write each simplified sum below.

7. $2x^2 + 3x^2$

   ____

8. $2x^2 + 3x + 4x^2 + 5x$

   ____

9. $4x^2 + 2x$

   ____

10. $2x^2 + 4x + 4 + x^2 + 2x + 2$

   ____

11. Multiplying with Algebra Tiles

Teacher Page

Learning Outcome

Students will be able to:

- multiply a monomial by a binomial using algebra tiles.
- multiply a binomial by a binomial using algebra tiles.
- discover multiplication properties of monomials and binomials.
- connect the physical model to an algebraic model.

Overview

Student pairs will use algebra tiles to multiply monomials and binomials, forming rectangles. They will also draw the products and express them algebraically with variables.

Time

45 minutes

Team Size

Pairs

Materials

Algebra tile sets for each team that have at least 6 x^2-tiles, 10 x-tiles, 12 unit tiles, and a separator bar

Procedure

Activity One

1. This activity deals with multiplying a monomial by a binomial. Prior to this activity, students should have discussed terms such as **monomial**, **binomial**, **trinomial**, and **polynomial**. They should also have some familiarity with the distributive property.
2. You should emphasize that the "factors" are the dimensions of a rectangle (or square) and that the "product" is the resulting area of the rectangle. (Note: this is true if the two or more factors produce a quadratic expression in which the highest degree is 2.)

Answers

Note that the placement of the dimensions is interchangeable. That is, the first factor can appear on the side or on top. Emphasize that this illustrates the **commutative property**.

1. 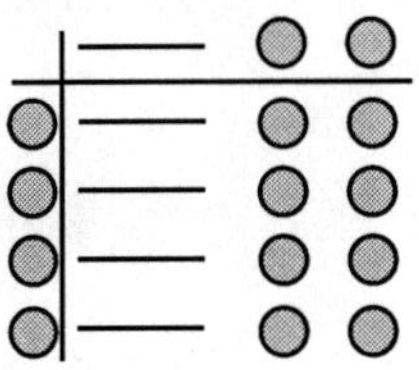

$4(x + 2) = 4x + 8$

(continued)

2. 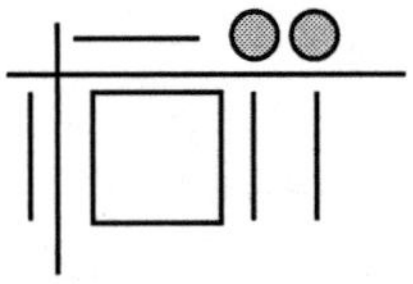

Students may have trouble at first determining which pieces fit. Stress the dimensions of the space. That is, the corner needs a piece with dimensions x by x *or* x^2.

$x(x + 2) = x^2 + 2x$

3. 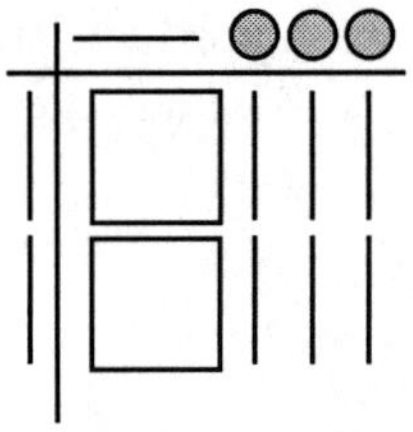

$2x(x + 3) = 2x^2 + 6x$

4. 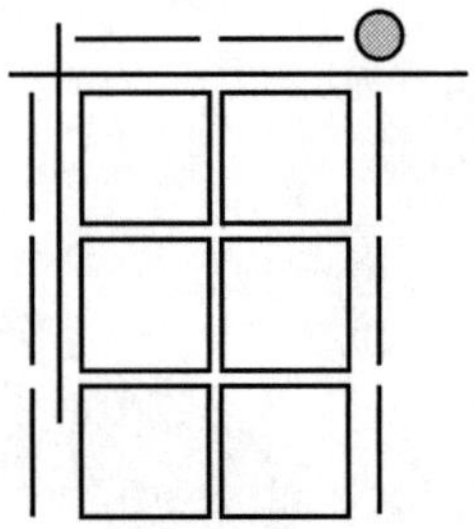

$3x(2x + 1) = 6x^2 + 3x$

Activity Two

1. In this activity, students should begin to notice some pattern relationships and placements with the tiles. For example, the squared terms are in the upper left corner, the x-tiles always surround the square tiles, and the constants (unit tiles) are always in the lower right corner. While there are other ways to arrange the tiles for the same results, this particular arrangement leads nicely into future discussions on squaring a binomial or completing the square.
2. In the final two questions for Activity Two, the first again verifies the commutative property.
3. The second question asks the students to look at the patterns evolving. A typical answer would be: The first term is formed by the product of the first terms in each binomial and the last term is formed by the product of the last terms in each binomial. Note that these two values are the upper left and lower right corners of the tiles. Finding the middle term of the product poses the most problems for students. Again note the placement of the x-tiles that produce the middle term of the product.
4. At this point, you may choose to discuss FOIL (First, Outer, Inner, Last), which the tiles clearly demonstrate.

(continued)

Answers

1.

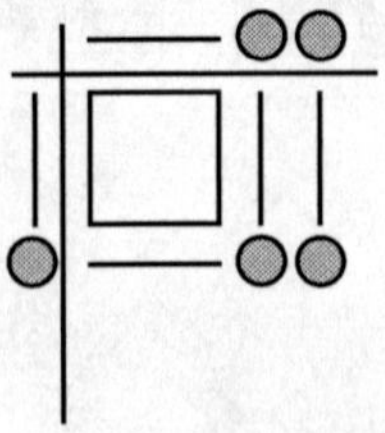

$(x + 1)(x + 2) = x^2 + 3x + 2$

2.

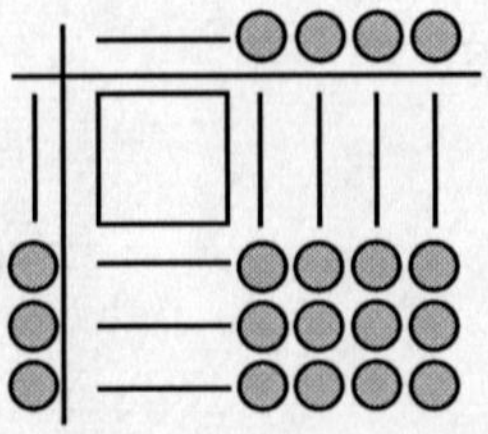

$(x + 3)(x + 4) = x^2 + 7x + 12$

3.

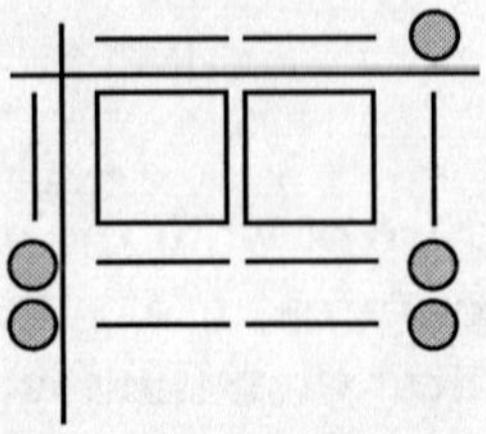

$(2x + 1)(x + 2) = 2x^2 + 5x + 2$

4.

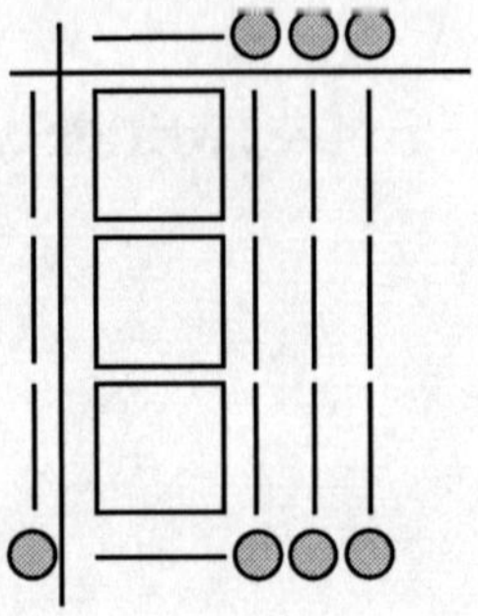

$(x + 3)(3x + 1) = 3x^2 + 10x + 3$

5.

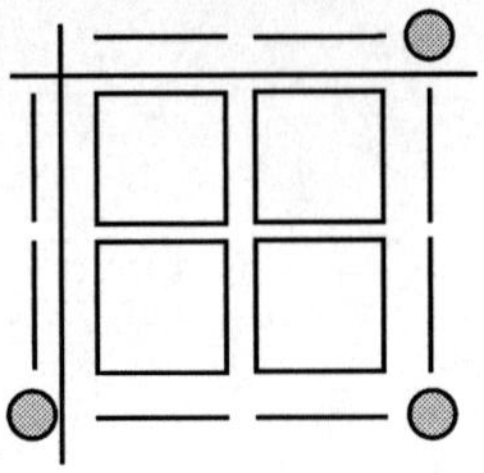

$(2x + 1)(2x + 1) = 4x^2 + 4x + 1$

Name ____________________ Date ____________

11. Multiplying with Algebra Tiles

Activity One

Using the gray separator that came with your tiles, make the rectangle whose dimensions are 2 by $x + 1$. In other words, $2(x + 1)$ will make what area that is rectangular in shape? In forming the inside **area** of the rectangle, make sure that tiles match correctly. In this first example the rectangle should look like:

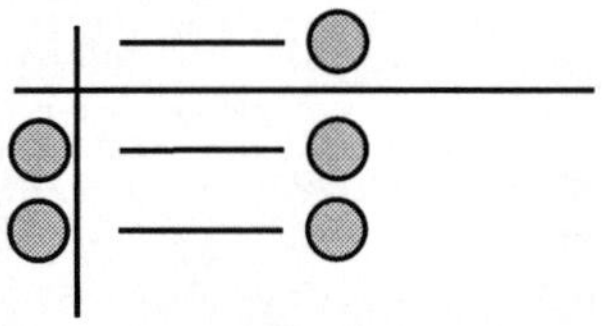

Therefore, $2(x + 1) = 2x + 2$.

Now find the following products with your tiles. Draw a picture of what the area of the rectangle looks like. Then algebraically define what the **product** (area) looks like with variables. Remembering the dimensions that are given, try to visualize what the shape will look like. For example, if you have dimensions ***x*** by ***x*** then the *x*-square tile should fit, if **1** by ***x*** then an *x*-tile will fit, and if **1** by **1** a unit tile should fit.

Drawing: **Product written algebraically:**

1. $4(x + 2)$

2. $x(x + 2)$

3. $2x(x + 3)$

Notice that each example above illustrates the **distributive property**. See if you can multiply the following algebraically, using the distributive property. Then verify your work using the algebra tiles.

(continued)

Name ____________________ Date ____________________

11. Multiplying with Algebra Tiles *(continued)*

4. $3x(2x + 1)$ Product written algebraically:

   Drawing:

Activity Two

We will now use the algebra tiles to find the product of a binomial and a binomial. Remember that each **factor** is the dimension of one side of the rectangle. Make a drawing of the end product.

	Drawing:	**Algebraic answer:**
1. $(x + 1)(x + 2)$		
2. $(x + 3)(x + 4)$		
3. $(2x + 1)(x + 2)$		
4. $(x + 3)(3x + 1)$		
5. $(2x + 1)(2x + 1)$		

Do you think $(x + 2)(x + 5)$ will give the same answer as $(x + 5)(x + 2)$? Verify with the tiles. Look at the answers to 1–5. Do you notice any patterns relating what you began with and what you ended up with?

Teacher Page

12. Factoring Trinomials with Algebra Tiles

Learning Outcome

Students will be able to:

- connect factoring of a quadratic to a rectangular model using algebra tiles.
- realize that only factorable trinomials of degree two will form rectangles.
- understand that "relatively prime" trinomials cannot be made into rectangles.
- understand that the trinomial given is the *area* of the rectangle and the factors found are the *dimensions* of the rectangle.
- discover strategies for factoring trinomials.

Overview

Students will use algebra tiles to factor trinomials. They will draw the algebra tiles on their worksheets and investigate the rules for factoring by looking at other teams' work along with their own.

Time

One hour

Team Size

Pairs

Materials

3 × 5 cards prepared with different pairs of binomials; sets of algebra tiles that contain x^2-tiles, x-tiles, and constant tiles. Each team should have at least 5 x^2-tiles, 10 x-tiles, and 10 constants.

Note: Students should have previous practice using algebra tiles (see the previous activities in this book).

Procedure

Activity One

1. Before beginning this activity, prepare 3 × 5 cards with different pairs of binomials to be multiplied, e.g., $(x + 1)\ (2x + 3)$. Make sure all binomials involve addition, not subtraction.
2. Pass out a 3 × 5 card to each group. Make sure each group gets a different pair of factors, and that each group has enough tiles to create the product.
3. Instruct each group to use the factors as dimensions of a rectangle and then to fill in the area with the correct matching tiles. You may wish to model the procedure for students.

Example: The team is given a card with $(x + 3)(2x + 4)$. Students place the dimensions on the outside of their separator bar.

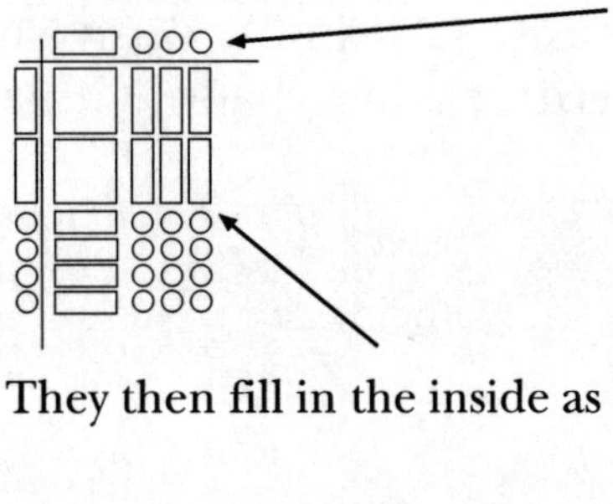

They then fill in the inside as above.

(continued)

4. Next, each group removes the outside dimensions and separator bar, leaving the interior rectangle.
5. Each group visits the other groups' tables and tries to figure out the dimensions of each rectangle and list its area as well.

   Example: The area of the rectangle above is $2x^2 + 10x + 12$. The dimensions should be $(x + 3)$ and $(2x + 4)$.
6. Then have them multiply the dimensions to see if they arrive at the area they observed.
7. The purpose of this first activity is for students to be able to recognize dimensions from the rectangle formed and the position of the tiles. Have them notice that the squared terms should be in the upper left corner, the x-tiles should surround the squared tiles, and the constants should appear in the lower right-hand corner. While other configurations are possible, having the students place the tiles in this manner allows for transferability in investigating rules for factoring and in completing the square later on.
8. Once all groups have had a chance to visit other tables, have each group report its rectangle's factors so that others can check to see if they guessed correctly.

Activity Two

1. Building upon the students' investigations in the first activity, you will now demonstrate how to factor using the algebra tiles.
2. The first example is $x^2 + 4x + 4$. Emphasize that the x^2 tiles should be placed in the upper left corner.
3. Demonstrate the various ways the x-tiles could be placed. Note that the x-tiles must border both sides of the x^2 tile. Possibilities are:

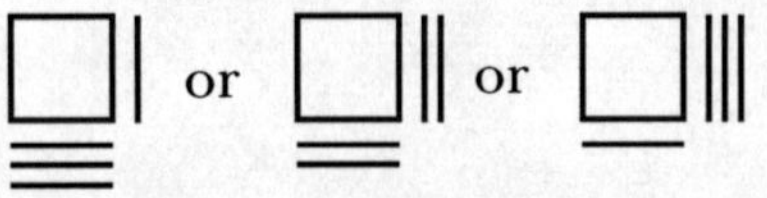

4. Tell students that the last term in the trinomial determines the correct choice. That is, 4 will only fit in the 2×2 space: 

   The resulting factors are: $(x + 2)(x + 2)$
5. In the second example, $2x^2 + 5x + 2$, the correct setup is:

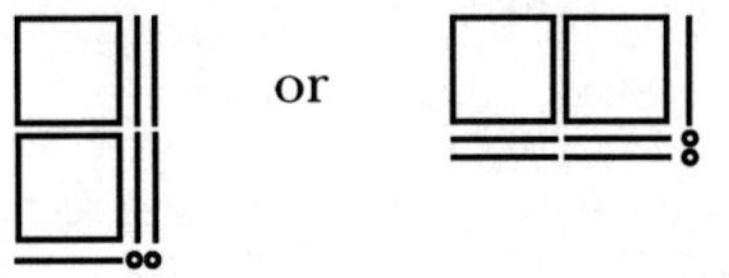

The resulting factors (dimensions) are $(2x + 1)(x + 2)$. Note that the commutative property is also illustrated with the tiles.

(continued)

6. The students now try several examples for practice.

Answers

1.

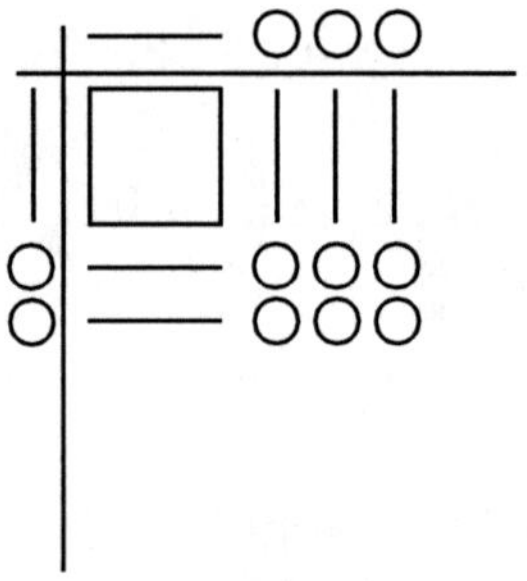

$x^2 + 5x + 6 = (x + 2)(x + 3)$

2.

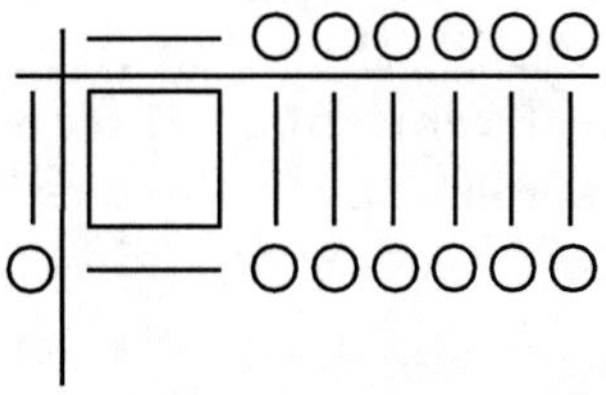

$x^2 + 7x + 6 = (x + 6)(x + 1)$

3.

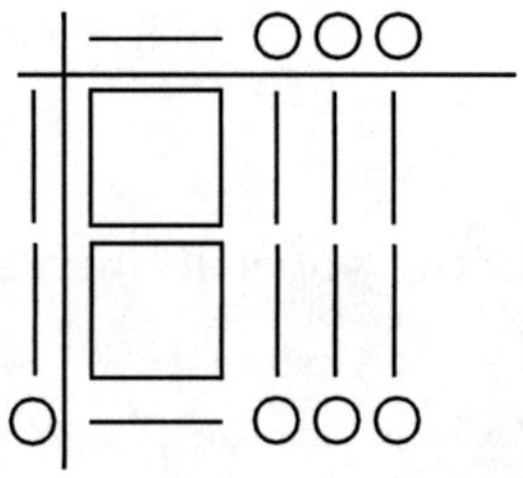

$2x^2 + 7x + 3 = (2x + 1)(x + 3)$

4.

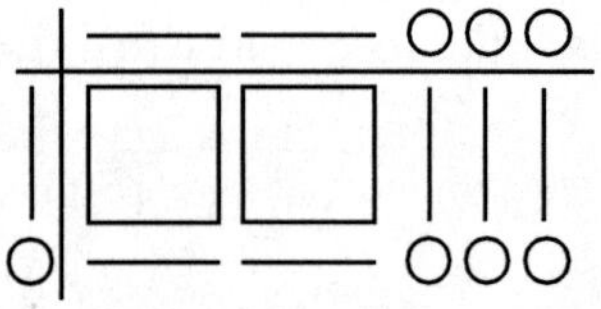

$2x^2 + 5x + 3 = (2x + 3)(x + 1)$

5.

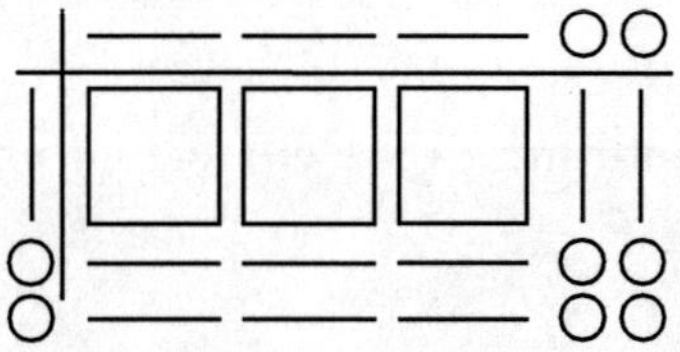

$3x^2 + 8x + 4 = (3x + 2)(x + 2)$

6.

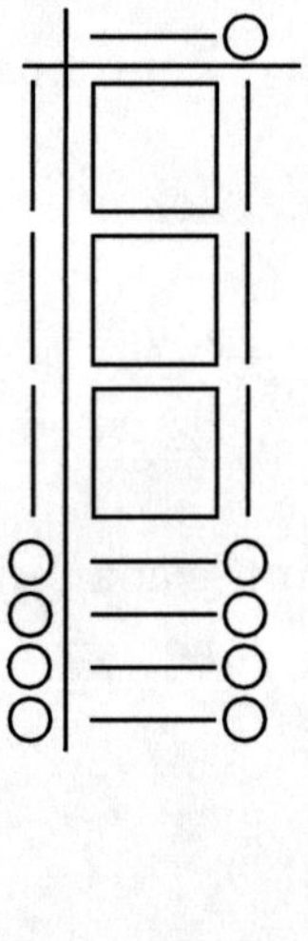

$3x^2 + 7x + 4 = (x + 1)(3x + 4)$

(continued)

7.

$3x^2 + 13x + 4 = (x + 4)(3x + 1)$

8.

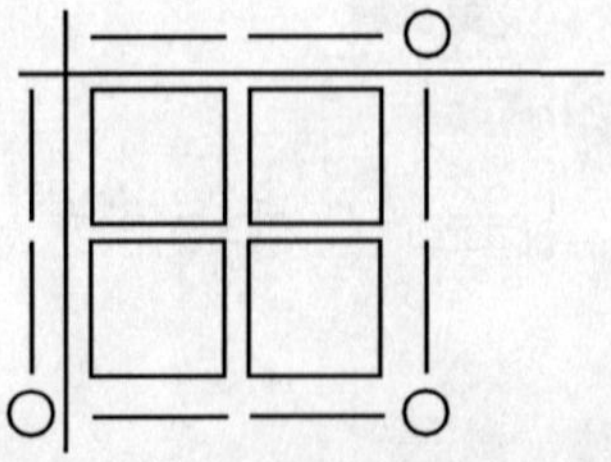

$4x^2 + 4x + 1 = (2x + 1)(2x + 1)$

9.

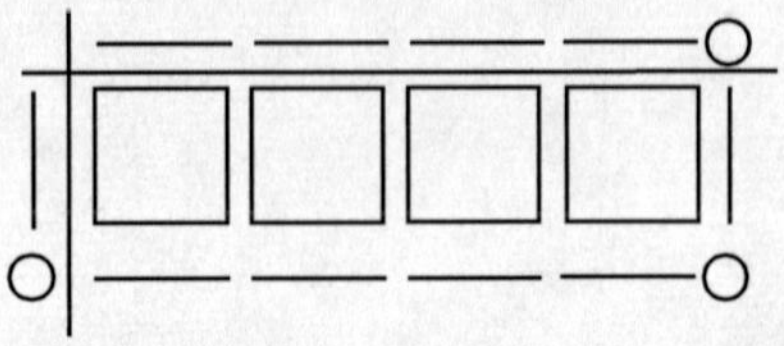

$4x^2 + 5x + 1 = (x + 1)(4x + 1)$

When students have completed all nine examples, ask them to look for any type of patterns. They may notice how the last term influences the middle term and how the tiles are arranged. They may also describe how the first term influences the placement of tiles, especially for coefficients of x^2 larger than 1.

Name ______________________________ Date ______________

12. Factoring Trinomials with Algebra Tiles

Activity One

Pick a card from the pile. The card will have the product of two binomials on it. With your tiles, multiply this to form the appropriate rectangle.

Sketch the rectangle below. Remove the separator piece and the two binomials from the table, leaving only the rectangle in place.

Once every team has completed the product on its card, visit the other tables and see if you can figure out what the two binomials are that are being multiplied. To do this, look at the rectangle formed and find its dimensions. The dimensions of the rectangle are the binomials you are looking for. Jot your guesses on the chart. Check your guess by multiplying the dimensions (binomials). The answer should be the area of the inside of the rectangle.

Table #	Two Binomials Being Multiplied	Resulting Product
	( )( ) =	
	( )( ) =	
	( )( ) =	
	( )( ) =	
	( )( ) =	

(continued)

Name ______________________________ Date ______________

12. Factoring Trinomials with Algebra Tiles *(continued)*

Activity Two

Now you are going to work backward. Your teacher is going to give you the area of a rectangle. Using the tiles, you are going to try to form and find the dimensions of that rectangle. In other words, you are going to **factor** the trinomial by changing the **sum** into the **product** of two binomials. If you find you cannot form a rectangle, then the sum cannot be factored and is said to be **prime**. Your teacher will do the first two examples with you.

Example 1:

Factor $x^2 + 4x + 4$ by finding the tiles that represent this sum and trying to form a rectangle. Sketch your rectangle and list the dimensions, the **factors**, of this sum below.

Example 2:

Factor $2x^2 + 5x + 2$

Now with your partner try factoring the following examples using your algebra tiles. Write your answer on the sheet. Multiply your answer to see if it is the same as the trinomial you started with. Remember to use the FOIL shortcut for multiplying binomials: **F**irst, **O**uter, **I**nner, **L**ast.

(continued)

Name ______________________________ Date ______________

12. Factoring Trinomials with Algebra Tiles *(continued)*

Trinomials	Factors	FOIL back to check
$x^2 + 5x + 6$		
$x^2 + 7x + 6$		
$2x^2 + 7x + 3$		
$2x^2 + 5x + 3$		
$3x^2 + 8x + 4$		
$3x^2 + 7x + 4$		
$3x^2 + 13x + 4$		
$4x^2 + 4x + 1$		
$4x^2 + 5x + 1$		

Can you and your partner see any patterns? List them below. (Hint: Look at the first and last term, see how they relate to the middle term. Also look at the rectangles formed.)

____

____

____

____

____

13. Perfect Square Trinomials

Teacher Page

Learning Outcome

Students will be able to:

- connect the geometric model of a perfect square to the algebraic model.
- recognize the various characteristics of a perfect square trinomial by observing the placement of algebra tiles.

Overview

Teams will use algebra tiles and observation to study geometric and algebraic models of perfect squares.

Time

30–45 minutes

Team Size

Pairs

Materials

Algebra tiles for each team; prepared 3×5 cards with various perfect square factors

Procedure

1. The purpose of this lab is for students to discover the various characteristics of a perfect square trinomial.
2. Before beginning this activity, prepare 3×5 cards with different pairs of perfect square binomials to be multiplied, e.g., $(x + 3)\ (x + 3)$, $(2x + 1)\ (2x + 1)$. Make sure that all binomials involve addition, not subtraction. Make enough cards that each group has a different product to find.
3. Give each group a 3×5 card.
4. Have the students build the rectangles that would be formed given the factors as dimensions. Students should have experience with multiplying algebra tiles. If not, a brief lesson on multiplying trinomials with algebra tiles prior to this lab will be necessary.
5. Once students have built their rectangles, have them remove the dimensions and leave only the rectangle in place.
6. Have students walk around the room and observe the other groups' rectangles (squares) and make observations to share.

Answers

1. Students should observe that the square tiles are grouped together, usually in the upper left-hand corner of the rectangle, and that the squares form a square within themselves.
   For example, suppose you gave out a card that read: $(2x + 3)\,(2x + 3)$

(continued)

It results in the following rectangle:

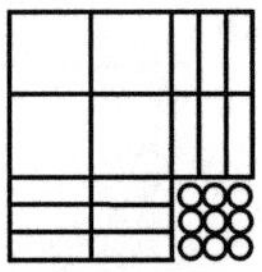

Note that the x^2 tiles form a square in the upper left-hand corner.

2. Students should notice that there is an even number of x-tiles bordering the edges of the x^2 tiles. In the example above, there are 6 x-tiles on either side of the x^2 square.
3. The students should notice that the unit tiles also form a square in the lower right-hand corner.
4. They are called this because the figure formed is a square.
5. Hopefully, with some observation and guidance, students will see that $(a + b)^2$ will result in $\mathbf{a^2}$ (the square in the upper left-hand corner) + $\mathbf{2ab}$ (doubling because the x-tiles are evenly distributed on either side of the a^2 section) + $\mathbf{b^2}$ (the resulting square of the unit tiles).
6. This example asks students to stretch a little further by giving them the interior of the rectangle (square) and asking them to find the dimensions. Hopefully, through tile manipulation, they will see that the first term, $4x^2$, would yield $2x$ for each dimension (that is, a $2x$ by $2x$ square), and that the last term would yield 3 as a dimension (that is, a 3 by 3 square). The factorization would be $(2x + 3)^2$ **Note:** You could extend this discovery by talking about completing the square using tiles. The unit tiles are missing and the student must determine how to complete the square, given that the other parts of the trinomial are known.

Name ______________________________ Date ______________

13. Perfect Square Trinomials

With your partner, draw a card from the pile. Create the rectangle with the tiles from the dimensions given on the card. When you are done, move about the room and observe the other tables' rectangles. Then answer the questions below.

1. What observations do you make about the square tiles? ______________________________
____
____

2. What observations do you make about the placement of the x-tiles? ______________
____
____

3. What observations do you make about the shape of the unit tiles? ______________
____
____

4. The rectangles you have formed are called **perfect square trinomials**. Why do you think they are called this? ______________________________
____
____

5. Given a binomial to square, that is $(a + b)(a + b)$ or $(a + b)^2$, can you see a short-cut to find the answer without using FOIL (First, Outside, Inside, Last)? (Hint: Remember what you noticed about the tiles.) ______________________________
____
____

6. $4x^2 + 12x + 9$ is a perfect square trinomial. What characteristics make it so?
____
____

14. Patterns and Slope-Intercept

Teacher Page

Learning Outcome

Students will be able to:

- determine if data will be linear.
- determine the formula for linear data using the slope-intercept formula.
- recognize, from a concrete pattern, what represents the constant value in the formula, and what represents the rate of change in the formula.

Overview

Groups will use square tiles to build three types of structures to illustrate the concepts of linear data and the slope-intercept formula. Students will enter their data in tables and devise formulas to describe each problem.

Time

1 hour

Team Size

2–3 students

Materials

Square tiles, preferably in two colors, at least 20–25 tiles per group

Procedure

Expanding Rectangles

1. The table should be filled in as follows:

Width x	Perimeter y
1	8
2	10
3	12
4	14

2. Stress that one way to determine whether data will be linear is to simply graph them. However, students can also recognize if data are linear by simply looking at the table. The input data, the ***x*** values, are in sequential order. The ***y*** values appear to increase in a constant pattern of 2. Tell students that when this occurs, the graphed data will be linear.
3. The constant **rate** the y values increase by is the **slope**.
4. To find the ***y*-intercept** from the table, simply follow the sequence back until $x = 0$. In this table, the student would have to go back one interval. That is, when $x = 0$, then $y = 6$. Note that in the physical world, it would be impossible to have a perimeter of 6 if there were no width, unless you consider the two lengths to be collapsed onto each other.
5. The real point in this lesson is to connect the abstract formula to the physical problem it is describing. In Expanding Rectangles, the width keeps changing while the

(continued)

length stays constant. The formula for expanding rectangles is $y = 2x + 6$. The slope of 2 represents the widths that keep increasing while the x represents how many stacks of widths there are. The y-intercept of 6 represents the 2 lengths (3 + 3) that do not change as the width increases.

Pigs in the Pen

Answers

1. The table would be as follows:

Pigs	Perimeter of pen
1	12
2	14
3	16
4	18

2. These data will be linear, as the perimeter increases by a constant of 2. In slope-intercept form, the formula would be $\mathbf{y = 2x + 10}$. The physical model can verify this, as each end contributes 5 edges to the perimeter and the inside pen increases by 2 each time. Look at the diagram below:

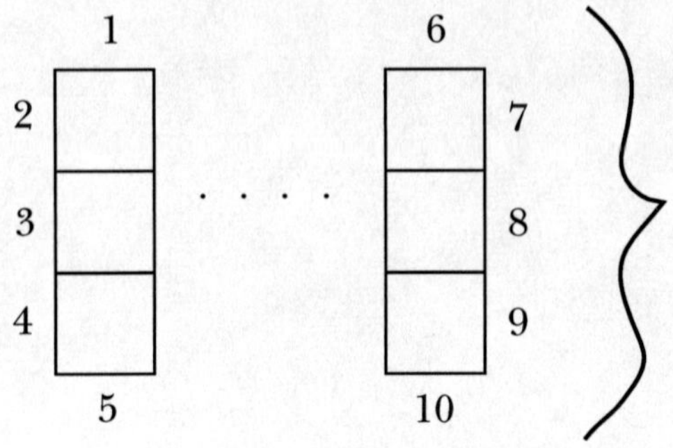

The outside edges of the pen's perimeter

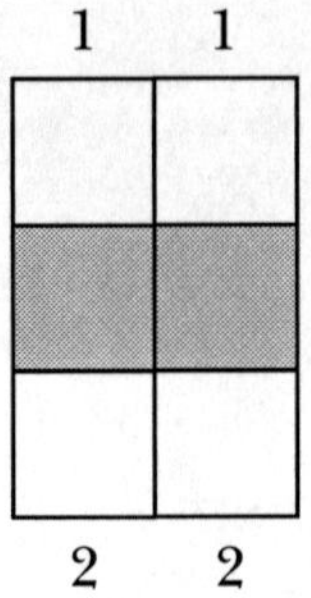

The inside sections of the pen's perimeter

Two edges are added each time a pig is included.
These edges keep expanding or changing by 2 each time.

3. The graph will cross the y-axis at 10. The table can also verify what the slope and y-intercept are. Note that the **y values** increase by a constant of 2, which is the constant rate or **slope**. If we extend the pattern back to when $x = 0$, we see that $x = 0$ and $y = 10$, which is the **y-intercept**.

(continued)

Barbecue Grill

1. The table values generated are:

Number of smokestack blocks	Total number of blocks
1	6
2	7
3	8
4	9

2. This pattern is described by the formula $y = 1 \bullet x + 5$. Yes, the formula is linear.
3. The graph will cross the y-axis at 5.
4. In this model, the **constant rate** or **slope** is one that we can verify with the table of values. Note that the y values increase by one each time. It can also be verified by the physical model. Note that the smokestack keeps changing by one cube each time.
5. The constant value is 5, which is the ***y*-intercept**.

Name ______________________________ Date ____________________

14. Patterns and Slope-Intercept

Expanding Rectangles

Take 3 tiles and place them side by side. Find the perimeter of the rectangle you formed and record it in the table below.

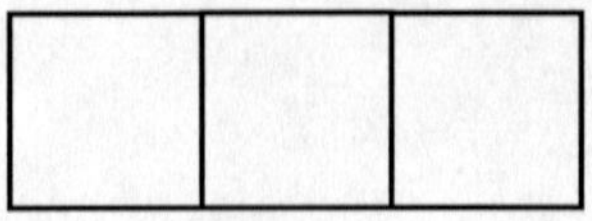

Width X	Perimeter Y

Now fit 3 more tiles under the first blocks and find the perimeter of the new rectangle. Put the results in the table above.

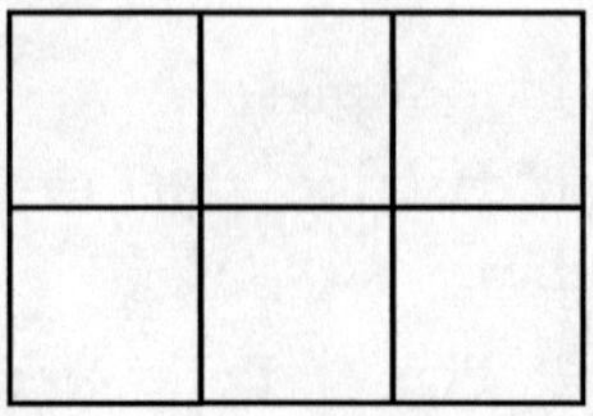

Repeat this pattern twice more, placing 3 more tiles underneath each time. Find each perimeter and put the values in the table.

1. Look carefully at the pattern that was created. What remained constant and what changed as you added tiles?
2. Can you find a formula to describe the pattern? Remember, what remained constant will be the **constant** in your formula and what changed will be the **rate** or **slope** for your formula.
3. Graph the data. Did the data form a straight line? ________
4. Look at your formula. Is this formula a linear equation? How do you know? Can it be written in slope-intercept form? ______________________________

____

____

(continued)

Name ______________________________ Date ______________

14. Patterns and Slope-Intercept *(continued)*

Pigs in the Pen

A farmer needs to build a pen to fence in his pig. For one pig the drawing would look like this:

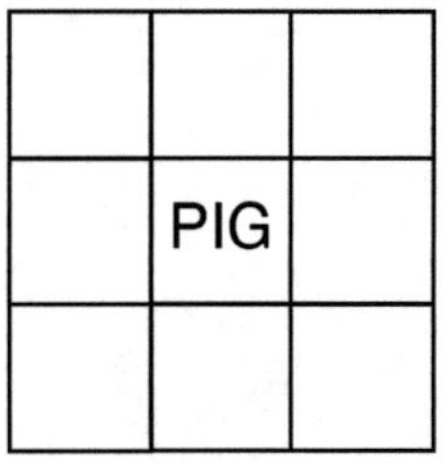

For two pigs:

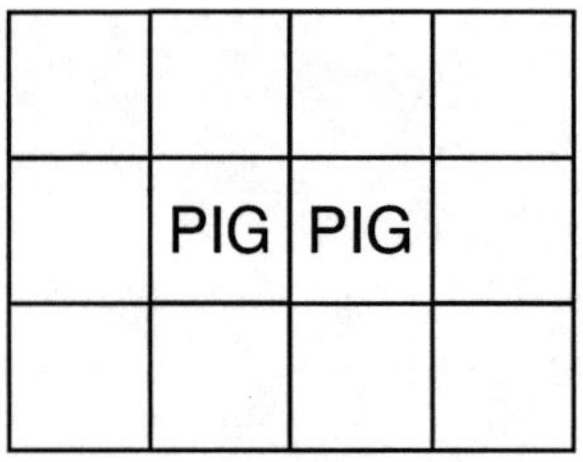

Continue building the pattern for three and four pigs.

1. Fill in the table below. See if you can "describe" the formula in terms of pigs (P) and the outer perimeter of fence cubes.

Formula:

Pigs	**Perimeter of Pen**

2. Is this formula linear? __________
   - Put the formula in slope-intercept form:

3. Where will the graph cross the *y*-axis? ______________________________

(continued)

Name ______________________________ Date ______________

14. Patterns and Slope-Intercept *(continued)*

4. How does the formula relate to the transformation of the blocks? (What remains constant and what changes with the blocks each time?)

____

____

5. How does the information in question 4 relate to slope and *y*-intercept?

____

____

Barbecue Grill

In this activity you are building a smokestack on a barbecue grill that has a base of 5 cubes with the following pattern:

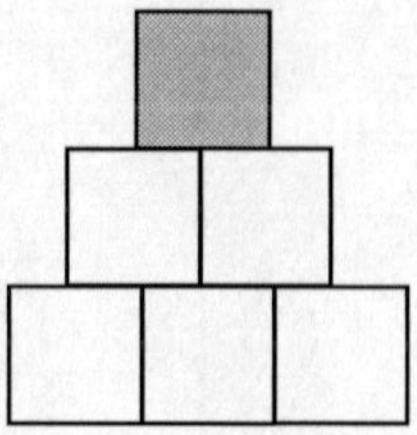

In the model above there is a 1-unit smokestack and a total of 6 blocks.

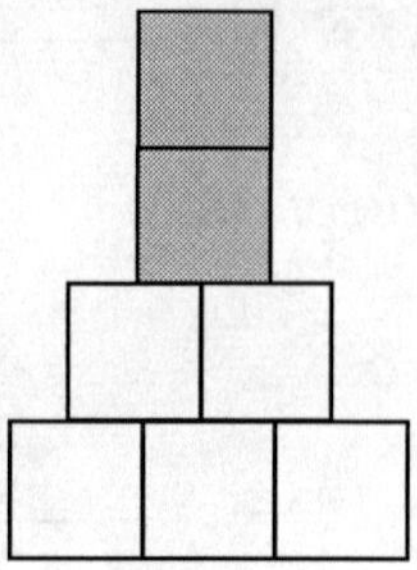

The model above shows a 2-unit smokestack and a total of 7 blocks.

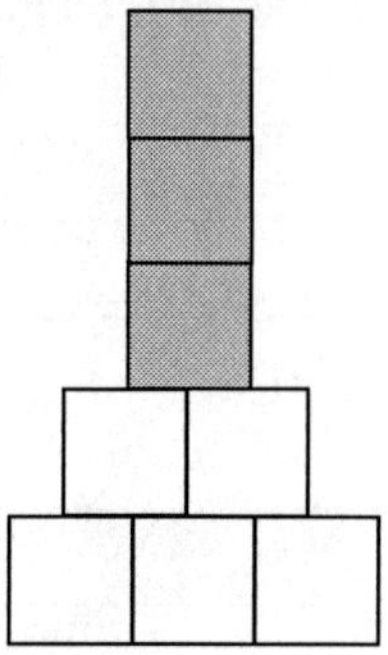

The model above has a 3-unit smokestack and a total of 8 blocks.

(continued)

Name ______________________________ Date ______________

14. Patterns and Slope-Intercept *(continued)*

1. The pattern continues. Fill in the chart below:

Number of smokestack blocks	**Total number of blocks**

2. What formula describes this pattern? ______________________________

   Is this formula linear? __________

   Put the formula in slope-intercept form:

3. Where will the graph cross the *y*-axis? ______________________________
4. How does the formula relate to the transformation of the blocks? (What remains constant and what changes with the blocks each time?)

   ____

   ____

   ____

5. How does the information in question 4 relate to slope and *y*-intercept?

   ____

   ____

   ____

Teacher Page

15. Slope as Rate

Learning Outcome

Students will be able to:

- recognize that rate and slope are the same quantity.
- calculate the slope or rate for data from a table or graph.
- determine the greater rate from a graph of several lines.
- express slope as rate, such as miles per hour, cost per employee, etc.

Overview

Teams will create data tables and graph the results, then answer questions that concern slope as rate, and rate of change.

Time

Approximately 90 minutes

Team Size

2–3 students

Materials

Graph paper

Procedure

Activity One: Slope as Rate

1. The purpose of this lab is for students to realize that slope is more than just plotting points on a line and determining the steepness. Rather, slope is a way of studying data and looking for how slowly or quickly data will change. Also, for linear functions, this change is constant and predictable, unlike curves, in which the rate keeps changing.
2. The first four situations concern the idea that the rate of change is constant. First the student must find that rate of change and then determine (interpolate) the outcomes between two given values.

Answers

CD example: The table is already filled in. Encourage students to look at intervals on the table and notice the "spread of data" between two given values to find the rate of change. For example, the spread between a $12 purchase and a $36 purchase is a gain of $24 (see diagram). This correlates to a spread of 1 CD to 3 CDs or 2 CDs purchased. The rate of change of dollars to CDs would be $24/2 or $12/1. Illustrate this for other intervals ($12 to $48 and 1 to 4) to show that this rate is constant at any interval.

CD	Cost
1	$12
2	$24
3	$36
4	$48

Spread of 2 (CDs 1 to 3) — Spread of $24 ($12 to $36)

(continued)

The rate is 24/2 or 12/1.

Students should realize that slope and rate are the same quantity.

1.

Hour	Degrees
1	–2
	+2.25
2	**.25**
	+2.25
3	**2.5**
	+2.25
4	**4.75**
	+2.25
5	7

The spread in degrees is 9 and the spread of hours is 4, so the rate of change in degrees per hour is 9/4 or 2.25.

2.

Hours	Miles
3	150
	+45
4	**195**
	+45
5	**240**
	+45
6	**285**
	+45
7	**330**
	+45
8	**375**
	+45
9	420

The spread of miles is 270 and the spread of hours is 6, so the rate of change in miles per hour is 45 mph.

3.

Hours	Miles
2	24
	+6
3	**30**
	+6
4	**36**
	+6
5	42

The spread of miles is 18 and the spread of hours is 3, so the rate of change in miles per hour is 6 mph.

Activity Two: Rate as a Graph

1. This activity has students examine the graphs more closely. The intent is for students to understand that slope tells us much about how slowly or rapidly change is occurring.

Answers

1. linear/nonlinear
   CD graph: linear; because the rate of change (slope) remained constant
   Degree change: linear; because the rate of change (slope) remained constant

(continued)

Auto trip: linear; because the rate of change (slope) remained constant
Bicycle trip: linear; because the rate of change (slope) remained constant

2. slope
CD graph: Here the student is asked to draw a right angle anywhere on the graph. The student should choose a right triangle that has whole number sides, however. The slope (12) will be the same as the rate of change.
Degree change: The right triangle students choose may not give the slope exactly. The correct response is 2.25, but students may only find 2.
Auto trip: The right triangle students choose may not give the slope exactly. The correct response is 45, but students may say 50.
Bicycle trip: They should find that the rate of change and the slope (6) are the same.

3. *y*-intercept:
CD graph: *y*-intercept is 0
Degree change: *y*-intercept is –4.25
Auto trip: *y*-intercept is 15
Bicycle trip: *y*-intercept is 12

4. equation in slope-intercept form
CD graph: $y = 12x + 0$
Degree change: $y = 2.25x - 4.25$
Auto trip: $y = 45x + 15$
Bicycle trip: $y = 6x + 12$

5. predicted value
CD graph: $y = 12(12) = \$144$
Degree change: $y = 2.25(9) - 4.25 = 16°$
Auto trip: $y = 45(10) + 15 = 465$ miles

6. The graph is based on data starting at 2 hours. The *y*-intercept indicates that they must have been going at a different rate for the first 2 hours. Looking at the table and the value (2,24) and comparing this to the ordered pair (0,0) for the start of the trip, the calculated slope would be 12 miles per hour instead of the 6 mph that was calculated with the data. This means that the bicyclists were traveling faster during the first two hours and slower the next 3 hours. The actual graph of the trip would look like this:

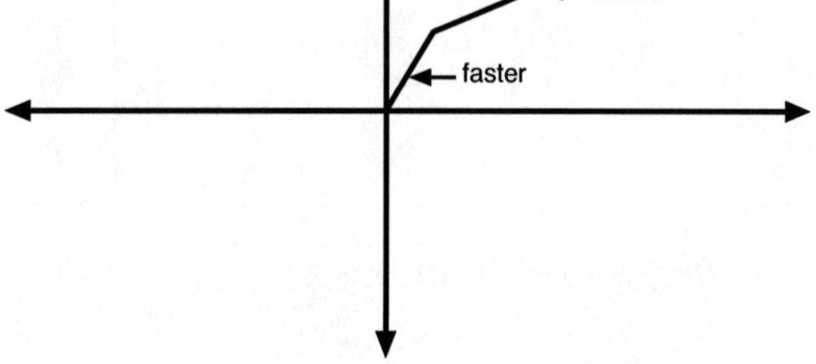

Name ______________________________ Date ______________________________

15. Slope as Rate

Activity One: Slope as Rate

Because lines are models of relationships between real-life quantities, slopes have real-world meanings. One of these meanings is related to the **rate of change** of quantities.

In this lab, you will investigate how two quantities relate to one another. One relationship to investigate is the **rate of change**. Examine the following table of data. Here you will compare the rate of change as **cost per CD**. To find the rate of change, divide the **cost** by the **number of CDs purchased**.

No. of CDs purchased	Cost of purchase
1	\$12
2	\$24
3	\$36
4	\$48

Graph the data. How does the **slope** compare to the **rate of change**?

Without graphing, find the rate of change in each situation below. Complete the table to help you visualize the problem.

1. Temperature readings are taken periodically in a 24-hour period. In the first hour, the temperature reading is –2 degrees. Four hours later, it is 7 degrees. What is the **rate of change** in **degrees per hour**? To answer this question, first determine the ratio of the change in degrees to the change in hours. Then fill in the table to verify:

Hour	Degrees
1	–2
2	
3	
4	
5	7

Rate of change in degrees per hour = __________

(continued)

Name ______________________________ Date ______________

15. Slope as Rate *(continued)*

2. An auto that has gone 150 miles in the first 3 hours of a trip continues on and goes a total of 420 miles in 9 hours. At what rate did the distance traveled change during the last **6 hours**? Determine the **rate of change** in **miles per hour** and fill in the table.

Hours	**Miles**
3	150
4	
5	
6	
7	
8	
9	420

Rate of change in miles per hour: __________

3. The cycle club had biked 24 miles after 2 hours and 42 miles after 5 hours. What was their **rate of change** in the last 3 hours expressed in **miles per hour**? Determine the rate in miles per hour and fill in the table below.

Hours	**Miles**
2	24
3	
4	
5	42

Rate of change in miles per hour: __________

(continued)

Name ______________________ Date ______________

15. Slope as Rate *(continued)*

Activity Two: Rate as a Graph

Look back at the four tables in Activity One. Plot each set of points on different graphs. Then answer the questions below.

1. Was the graph linear? What makes it linear?

   CD graph: ______________________

   ____

   Degree change: ______________________

   ____

   Auto trip: ______________________

   ____

   Bicycle trip: ______________________

   ____

2. Find the slope using a right triangle. How does it compare to the slope you calculated in Activity One?

   CD graph: ______________________

   Degree change: ______________________

   Auto trip: ______________________

   Bicycle trip: ______________________

3. Extend the line of the graph so that it intersects the y-axis. What is the y-intercept?

   CD graph: ______________________

   Degree change: ______________________

   Auto trip: ______________________

   Bicycle trip: ______________________

4. What is the equation (formula) of the line in slope-intercept form?

   CD graph: ______________________

   Degree change: ______________________

   Auto trip: ______________________

   Bicycle trip: ______________________

(continued)

Name ________________________________ Date ______________

15. Slope as Rate *(continued)*

5. You can use your formula to predict any value.

   CD graph: What would be the cost if you bought 12 CDs? ______________

   Degree change: What would the temperature be after 9 hours? ______________

   Auto trip: How far will you have traveled after 10 hours? ______________

6. Bicycle trip: When $x = 0$, or no hours, the distance traveled should be 0 as well. Yet the graph indicates they had traveled 12 miles. What caused this inconsistency?

   ____

   ____

   ____

   ____

Teacher Page

16. Graphing Curves with a Graphing Calculator

Learning Outcome

Students will be able to:

- determine how wide or how narrow a parabola will be.
- determine the location of a parabola on the *y*-axis.

Overview

Student teams will use a graphing calculator to make discoveries about the effects of the coefficient and the constant on the width and location, respectively, of a parabola.

Time

45 minutes

Team Size

Pairs

Materials

Graphing calculators (If graphing calculators are not available, this activity can be done using graph paper.)

Procedure

If graphing calculators are not available, direct students to plot the curves on graph paper.

Activity One

Answers

4. Students should discover the effect the coefficient of a squared term has on a graph. They should find that the coefficient has an effect on the broadness or narrowness of the parabola.
6. As the absolute value of the coefficient gets larger, the parabola gets skinnier.
7. As the coefficient's value is between 0 and 1, the parabola gets very wide.

Activity Two

1. As students graph the parabolas 8–10, they should observe that the shape of the parabola stays the same, but moves up and down on the *y*-axis. They should discover that the constant being added or subtracted determines where the vertex of the parabola will cross the *y*-axis.
2. Question 13 asks the student to graph a parabola that uses all the characteristics discovered in the lab. The graph of $y = 3x^2 - 2$ will result in a somewhat skinny parabola whose vertex is at $(0,-2)$.
3. In question 14, it is hoped the students see that the coefficient of *x* has an effect on both linear and nonlinear graphs. In linear graphs, it determines how steep or gradual a line will be, while in the nonlinear case, the coefficient will determine how skinny or broad the parabola will be. Also, the constant term determines the *y*-intercept, or place where the line or vertex of the parabola intersects the *y*-axis.

Extensions

You might have students explore the graphs of $y = (x - h)^2$ or $y = a(x - h)^2 + b$.

Name ______________________ Date ______________

16. Graphing Curves with a Graphing Calculator

Activity One

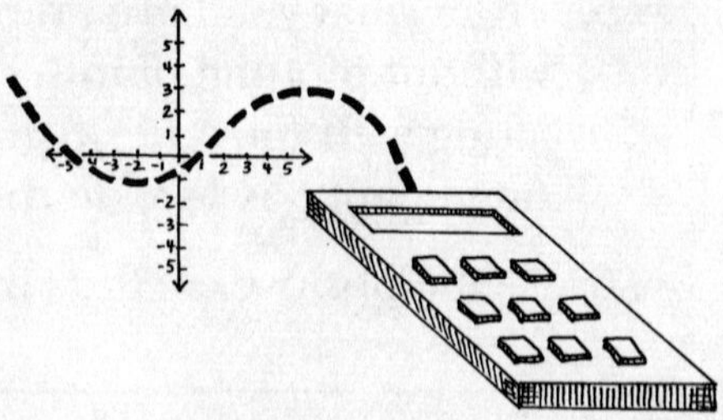

For the following equations (1–3), complete the table. Then graph the points on the same coordinate grid by using your graphing calculator.

1. $y = x^2$

x	y
0	
1	
–1	
2	
–2	
3	
–3	

2. $y = (1/2)x^2$

x	y
0	
2	
–2	
4	
–4	

3. $y = 2x^2$

x	y
0	
1	
–1	
2	
–2	

4. What do you notice about the shapes of these graphs?

____

____

(continued)

Name ______________________________ Date ______________

16. Graphing Curves with a Graphing Calculator *(continued)*

5. From what you have graphed, can you determine where the graph of $y = 4x^2$ will be without plotting points? __________ Check your assumption by graphing $y = 4x^2$ along with the other 3 graphs. Use your graphing calculator or plot points on the same paper as 1–3, on the previous page.

6. As the positive coefficients of x get larger, what happens to the shape of the graph?

   ____

7. As the positive coefficients of x get smaller, what happens to the shape of the graph?

   ____

Activity Two

For the following equations (8–10), complete the table and then graph the points on the same coordinate grid by either plotting the points or using your graphing calculator.

8. $y = x^2 + 1$

x	y
0	
1	
–1	
2	
–2	
3	
–3	

9. $y = x^2 + 3$

x	y
0	
1	
–1	
2	
–2	
3	
–3	

(continued)

Name ________________________________ Date ________________

16. Graphing Curves with a Graphing Calculator *(continued)*

10. $y = x^2 - 4$

x	y
0	
1	
–1	
2	
–2	
3	
–3	

11. Did the shape of the graph change? What did happen? ________________________________

____

____

12. With this information, could you sketch the graph of $y = x^2 + 5$ without plotting points? ________ Check your assumption by graphing $y = x^2 + 5$ along with graphs 8–10. Use your graphing calculator or plot points on the same paper as the other 3 graphs.

13. With the information from this entire lab could you sketch the following graph?
$y = 3x^2 - 2$
Verify with your graphing calculator or graph paper.

14. How does the graph of $y = ax^2 + b$ relate to $y = mx + b$? ________________________

____

____

17. How Long Is the Bounce?

Teacher Page

Learning Outcome

Students will be able to:

- make predictions and identify formulas from graphed data.
- determine how possible errors in data collection can affect the outcome of an experiment.

Overview

Groups will time the bouncing periods of tennis balls and Ping-Pong balls dropped from different heights, record the data, graph the curves, and make predictions.

Time

60–90 minutes

Team Size

3 or 4 students

Materials

Tennis ball, Ping-Pong ball, stop watch, tape measure, data sheet, calculator, and graph paper for each team; access to computer and spreadsheet program (optional)

Procedure

1. Before beginning the lab, have groups decide which members will perform which task. Duties include a timer, a data collector, and someone to hold and release the ball.
2. Tell students that they will first collect data on the tennis ball by conducting four trials at various heights and measuring how long the ball bounces.
3. Tell students that they should begin timing once the ball is released and stop timing when the ball no longer bounces.
4. Once the data are collected on the tennis ball, the students should average the trials, graph the averages, and discuss the various questions that follow *before* experimenting with the Ping-Pong ball.

Answers

1. The ordered pair (0,0) is necessary to fully identify the shape of the graph. This implies that there is no bounce when the ball is not raised to a height.
2. The relationship is nonlinear.
3. Usually students estimate that at the $\frac{1}{2}$-foot mark, the ball will bounce for approximately 1 minute. They arrive at that answer by looking at the graph.
4. The shape of the curve should resemble the graph for the equation of $y = \sqrt{x}$

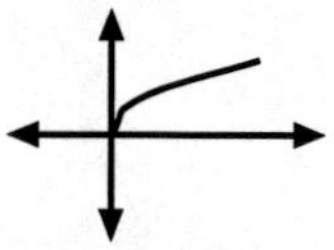

5–6. These questions allow students to reflect on and analyze what happened with the tennis ball and how the Ping-Pong ball will affect the shape of the curve. Many students will correctly determine that the shape of the curve will remain the same, but will go much higher. This part of the lab helps you observe how well students can interpret the cause-and-effect relationship of a graph.

Name ______________________________ Date ______________

17. How Long Is the Bounce?

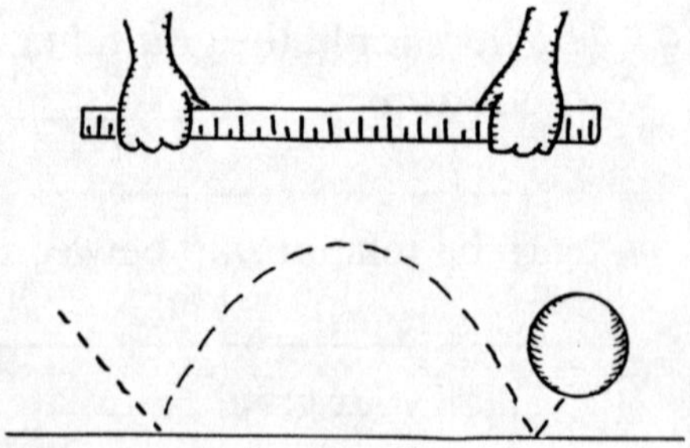

The purpose of this lab is to examine the length of time the ball bounces from different heights and to plot this data on a grid.

Drop the tennis ball onto the floor from various heights. At each designated height, time how long the ball bounces. Record this in the table below. Repeat each trial 4 times.

Drop height	Time of bounce (4 trials)			
6				
5				
4				
3				
2				
1				

After 4 trials, calculate the average time the ball bounces for each drop. Record the average time on the table below.

Drop height	Average time
6 ft.	
5 ft.	
4 ft.	
3 ft.	
2 ft.	
1 ft.	

Graph the data from this table on a piece of graph paper, or enter the data into a computer using a spreadsheet and chart command. Use the drop height as the independent variable (x-value) and time as the dependent variable (y-value).

(continued)

Name ______________________ Date ______________

17. How Long Is the Bounce? *(continued)*

1. Why should the graph begin at (0,0)? What does this imply about the drop height and bounce? ______________________

____

2. Is the relationship between drop height and bouncing time linear or nonlinear?

____

3. From your graph, estimate the time it would take the tennis ball to bounce at ½ foot. ____________ Now go back and time the ball at the ½ ft. mark. How close was your estimate to the actual timing? ______________________

4. The shape of the curve most resembles which of the following relationships:

   (a) $y = x^2$

   (b) $y = \sqrt{x}$

   (c) $y = 1/x$

   Repeat this experiment using a Ping-Pong ball. Before you begin, answer the following questions:

5. Do you think the results using the Ping-Pong ball will be the same? ____________

6. Discuss how you think the graph may change. Do you think the equation you chose in #4 will be the same for the Ping-Pong ball? ____________

   Begin collecting data.

Drop height	Time of bounce (4 trials)			
6				
5				
4				
3				
2				
1				

(continued)

Name ______________________________ Date ______________

17. How Long Is the Bounce? *(continued)*

Calculate the average time the ball bounces for each drop after 4 trials. Record the drop height and the average time on the table below.

Drop height	Average time
6 ft.	
5 ft.	
4 ft.	
3 ft.	
2 ft.	
1 ft.	

Graph the data from the last table on a piece of graph paper. Use the drop height as the independent variable (*x*-value) and time as the dependent (*y*-value) variable.

7. Were your predictions in questions 5 and 6 verified during the experiment, or did something happen that you did not expect? Explain.

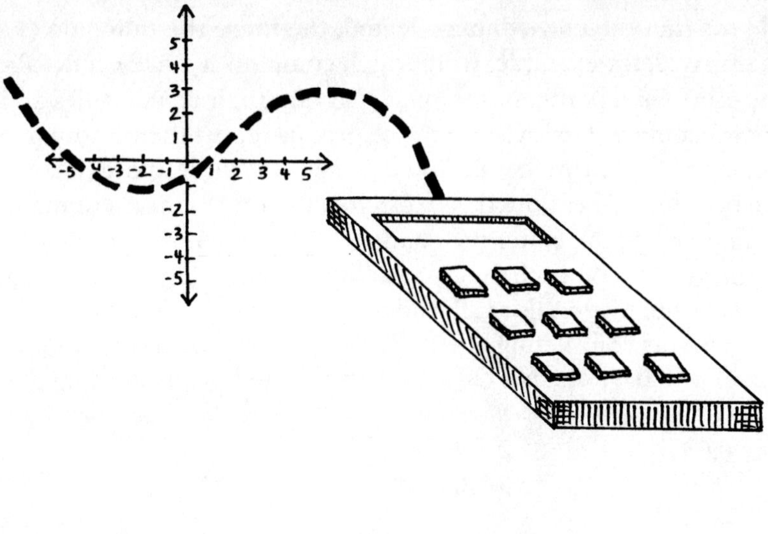

Teacher Page

18. Calendar Number Probability

Learning Outcome

Students will be able to:

- understand the concept of simple probability.
- understand how the size of a sample space can alter data outcomes.

Overview

Teams will experiment with probability by making predictions and then drawing from 30 numbers in a bag over 20 trials. Teams will pool data to see that the number of trials increases the consistency of the data.

Time

45 minutes

Team Size

Pairs

Materials

Numbers 1–30 for each group (a good source is old calendars); paper bags in which to store numbers

Procedure

1. Prior to the lab, cut out numbers 1–30 from used calendars, or print out sets of numbers on a computer. Put one set of numbers in each paper bag. Make enough sets to supply a bag of 30 numbers for each group in your classroom.
2. Have each team collect data on their drawings as suggested in the lab. They draw a number, tally it on the sheet, and return it to the bag.
3. Ask students which number came up the most when drawing 20 times, and whether this happened in each group.
4. Place a master grid like the one illustrated for the lab on the front board and have students post their tallies for the numbers they have drawn.
5. Students should see that the more data collected, the more the outcome changes. Discuss how this would affect surveys, for example, from which companies make claims about their products.
6. As students conduct their experiment they may find that their actual tallies do not come close to their predicted probabilities. However, when the groups report their findings and combine the results, the outcome should prove better. You may have to gather more data by running the experiment a second time. The more data collected, the more consistent the outcome.
7. Remember, the more trials, the better the chances of coming to the predicted probability. Be sure that, when the groups compute their probabilities, they compare the number of times the number was drawn to the total trials of all students. For example, say the number 13 was drawn 15 times with 20 students conducting the trial. The actual probability would be 15 out of 600 (20×30 numbers) or 0.025, which is close to the predicted probability of 0.03.

Answers

1. 1/30 or about 0.03
2. 15/30 or .50
3. 21/30 or .7
4. 9/30 or .3

Name ____ Date ______________________

18. Calendar Number Probability

Before experimenting with the data, answer the following questions. In your bag there are 30 squares numbered 1–30.

1. What is the probability that you will draw the number 13 from the bag? ______________
2. What is the probability that you will draw an even number from the bag? ______________
3. What is the probability that you will draw a two-digit number from the bag? ______________
4. What is the probability that you will draw a single-digit number from the bag? ______________

Draw a number from the container. Put a tally mark under the number on the chart each time you draw that number. Place the number **back** in the bag. Repeat this process **20** times.

1	2	3	4	5	6	7	8	9	10
11	12	13	14	15	16	17	18	19	20
21	22	23	24	25	26	27	28	29	30

Now share the results of your experiment with the rest of the class. How does increasing the data collected change the outcome?

____

____

____

____

19. Discovering Log Properties

Teacher Page

Learning Outcome

Students will be able to:

- derive the various properties for logarithms through a variety of experiments.

Overview

Pairs of students will use a scientific calculator and worksheet exercises to discover properties of logarithms.

Time

45 minutes

Team Size

Pairs

Materials

Scientific calculators

Procedure

1. Prior to this lab, students should be familiar with how to write a number in logarithmic setup to exponential setup. That is, $a^x = y$ can be written using logarithms as $\log_a y = x$.
2. Pass out the lab sheet and have students work through it to discover each property.

Answers

Activity One

1. $\log_2 2 = y = 1$
2. $\log_3 3 = y = 1$
3. $\log_4 4 = y = 1$
4. $\log_5 5 = y = 1$
5. $\log_a a = 1$
6. $\log_2 1 = y = 0$
7. $\log_3 1 = y = 0$
8. $\log_4 1 = y = 0$
9. $\log_5 1 = y = 0$
10. $\log_a 1 = 0$

(continued)

Activity Two

1. $\log_2 (4 \bullet 8) = \underline{5}$
2. $\log_3 (3 \bullet 9) = \underline{3}$
3. $\log_4 (16 \bullet 4) = \underline{3}$
4. $\log_2 4 + \log_2 8 = \underline{2 + 3 = 5}$
5. $\log_3 3 + \log_3 9 = \underline{1 + 2 = 3}$
6. $\log_4 16 + \log_4 4 = \underline{2 + 1 = 3}$
7. $\log_a xy = \log_a x + \log_a y$

Activity Three

1. $\log_2 (16 \div 2) = \underline{3}$
2. $\log_2 (32 \div 4) = \underline{3}$
3. $\log_3 (81 \div 27) = \underline{1}$
4. $\log_2 16 - \log_2 2 = \underline{4 - 1 = 3}$
5. $\log_2 32 - \log_2 4 = \underline{5 - 2 = 3}$
6. $\log_3 81 - \log_3 27 = \underline{4 - 3 = 1}$
7. $\log_a (x \div y) = \log_a x - \log_b x$

Activity Four

1. $\log_2 2^3 = \underline{3}$
2. $\log_3 9^2 = \underline{4}$
3. $\log_2 4^2 = \underline{4}$
4. $3 \bullet \log_2 2 = \underline{3 \bullet 1 = 3}$
5. $2 \bullet \log_3 9 = \underline{2 \bullet 2 = 4}$
6. $2 \bullet \log_2 4 = \underline{2 \bullet 2 = 4}$
7. $\log_a x^p = p \bullet \log_a x$

Activity Five

1. $\log_3 3^4 = \underline{4}$
2. $\log_2 2^5 = \underline{5}$
3. $\log_4 4^3 = \underline{3}$
4. $4 \bullet \log_3 3 = \underline{4 \bullet 1 = 4}$
5. $5 \bullet \log_2 2 = \underline{5 \bullet 1 = 5}$
6. $3 \bullet \log_4 4 = \underline{3 \bullet 1 = 3}$
7. $\log_a a^p = p$

Name ______________________________ Date ______________

19. Discovering Log Properties

Activity One

With your partner, find the values for *y* in the following:

1. $\log_2 2 = y$
2. $\log_3 3 = y$
3. $\log_4 4 = y$
4. $\log_5 5 = y$
5. Discuss what you noticed happening in each example. Finish the following general rule:

$$\log_a a = \_\_\_\_\_$$

With your partner, find the values for *y* in the following:

6. $\log_2 1 = y$
7. $\log_3 1 = y$
8. $\log_4 1 = y$
9. $\log_5 1 = y$
10. Discuss what you noticed happening in each example. Finish the following general rule:

$$\log_a 1 = \_\_\_\_\_$$

Activity Two

With your partner, try the following examples:

1. $\log_2 (4 \bullet 8)$ = __________
2. $\log_3 (3 \bullet 9)$ = __________
3. $\log_4 (16 \bullet 4)$ = __________
4. $\log_2 4 + \log_2 8$ = __________
5. $\log_3 3 + \log_3 9$ = __________
6. $\log_4 16 + \log_4 4$ = __________
7. Discuss what you noticed happening in each example. Finish the following general rule:

$$\log_a xy = \_\_\_\_\_\_\_\_\_\_$$

(continued)

Name ______________________________ Date ______________

19. Discovering Log Properties *(continued)*

Activity Three

With your partner, try the following examples:

1. $\log_2 (16 \div 2)$ = __________
2. $\log_2 (32 \div 4)$ = __________
3. $\log_3 (81 \div 27)$ = __________
4. $\log_2 16 - \log_2 2$ = __________
5. $\log_2 32 - \log_2 4$ = __________
6. $\log_3 81 - \log_3 27$ = __________
7. Discuss what you noticed happening in each example. Finish the following general rule:

$$\log_a(x \div y) = \_\_\_\_\_\_\_\_\_\_$$

Activity Four

With your partner, try the following examples:

1. $\log_2 2^3$ = __________
2. $\log_3 9^2$ = __________
3. $\log_2 4^2$ = __________
4. $3 \bullet \log_2 2$ = __________
5. $2 \bullet \log_3 9$ = __________
6. $2 \bullet \log_2 4$ = __________
7. Discuss what you noticed happening in each example. Finish the following general rule:

$$\log_a x^p = \_\_\_\_\_\_\_\_\_\_$$

Activity Five

With your partner, try the following examples:

1. $\log_3 3^4$ = __________
2. $\log_2 2^5$ = __________
3. $\log_4 4^3$ = __________
4. $4 \bullet \log_3 3$ = __________
5. $5 \bullet \log_2 2$ = __________
6. $3 \bullet \log_4 4$ = __________
7. Discuss what you noticed happening in each example. Finish the following general rule:

$$\log_a a^p = \_\_\_\_\_\_\_\_\_\_$$

Teacher Page

20. Math Project: Researching Internet Providers

Learning Outcome

Students will be able to:

- gather, organize, visualize, and analyze data.
- display data graphically.

Overview

Student teams will undertake a major research and reporting assignment. They will gather information on features and costs and report it in a variety of mathematical and verbal ways.

Time

2 weeks or more, as decided by teacher and students

Team Size

2 or more students

Materials

Optimal: poster board, markers, brochures from several Internet providers

Procedure

As students gather information, they are asked to present the material in a variety of ways. Possibilities include:

Tables: In a table, students could show the cost of various systems for the time used.

Graphs: Students might also construct linear graphs to show which company offers the best deal. The graphs could also show different companies offering better choices for a given amount of time on-line.

Formulas: Student teams could find formulas for calculating a customer's cost given a set of variables.

Surveys: The survey is another way of collecting data and then mathematically communicating the findings. Teams could show results as a frequency distribution, a histogram, or a bar graph.

Brochures: The handout also asks students to use a brochure to communicate their findings and to provide good arguments for their decision. Here they will be using math as reasoning.

This project could be adapted to long-distance telephone offers, car rental offers, car leasing, etc.

Name ______________________________ Date ______________

20. Math Project: Researching Internet Providers

Imagine . . . you have been hired by a consumer group to investigate the various Internet providers in your area and do a cost analysis. Many providers have appeared in the area offering a variety of packages.

Your task is to:

1. Find and research at least 4 different providers in your area.

   Provider 1: ______________________________

   Provider 2: ______________________________

   Provider 3: ______________________________

   Provider 4: ______________________________

2. Compare and contrast the various suppliers through:

   (a) developing tables of data for comparison purposes.

   (b) creating graphs.

   (c) developing formulas to calculate costs for the customer.

3. Develop a survey and question at least 50 people about what features and fees they seek from an Internet provider. Display this data visually.

4. Create a brochure to inform the public of your findings and recommendations. Back up your suggestions with reasons.

Share Your Bright Ideas

We want to hear from you!

Your name________________________________Date________________

School name____

School address____

City____________________State____Zip_______Phone number (____)________

Grade level(s) taught______Subject area(s) taught________________________

Where did you purchase this publication?______________________________

In what month do you purchase a majority of your supplements?____________________

What moneys were used to purchase this product?

___School supplemental budget ___Federal/state funding ___Personal

Please "grade" this Walch publication in the following areas:

Quality of service you received when purchasing	A	B	C	D
Ease of use	A	B	C	D
Quality of content	A	B	C	D
Page layout	A	B	C	D
Organization of material	A	B	C	D
Suitability for grade level	A	B	C	D
Instructional value	A	B	C	D

COMMENTS:____

____

What specific supplemental materials would help you meet your current—or future—instructional needs?

____

Have you used other Walch publications? If so, which ones?________________________

May we use your comments in upcoming communications? ___Yes ___No

Please **FAX** this completed form to **888-991-5755**, or mail it to

Customer Service, Walch Publishing, P. O. Box 658, Portland, ME 04104-0658

We will send you a **FREE GIFT** in appreciation of your feedback. **THANK YOU!**